THE YOUTH GUIDE TO THE OCEAN

1ST EDITION

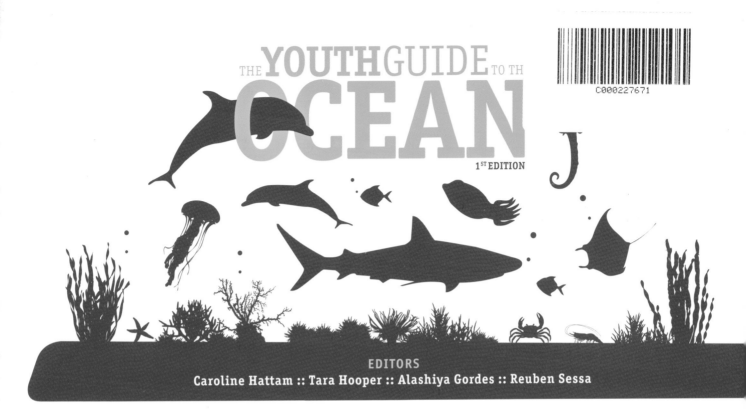

EDITORS

Caroline Hattam :: Tara Hooper :: Alashiya Gordes :: Reuben Sessa

AUTHORS

:: José Aguilar-Manjarrez :: David Billett :: Kelvin Boot :: Jennifer Corriero :: Kelly-Marie Davidson ::
:: Emily Donegan :: Annie Emery :: Helen Findlay :: Nicole Franz :: Alashiya Gordes :: Caroline Hattam ::
:: Tara Hooper :: Frances Hopkins :: Jennifer Lockett :: Alessandro Lovatelli :: Ana M. Queirós ::
:: Reuben Sessa :: Jack Sewell :: Doris Soto :: Jogeir Toppe :: Christi Turner ::

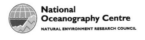

ISBN 978-92-5-108647-6

This document has been financed by the Swedish International Development Cooperation Agency, Sida. Sida does not necessarily share the views expressed in this material. Responsibility of its contents rests entirely with the authors.

FRONT COVER PHOTO:
© 4ever.eu

BACK COVER PHOTOS:
© Richard Shucksmith, Earth in Focus :: © Anton Bielousov, WMC :: © Pachango, Flickr :: © Ben Tubby, WMC :: © Efraimstochter, All-free-download.com

Acknowledgements

The Youth and United Nations Global Alliance (YUNGA) would like to thank all the authors, contributors, graphics designers and other individuals and institutions who have supported the development of this Guide. All have found the extra time in their busy schedules to write, edit, prepare or review the publication, and many others have kindly allowed the use of their photos or other resources. We are deeply grateful to Chris Gibb, Neil Pratt and Chantal Robichaud at CBD; Isabelle Brugnon, Rejane Herve-Smadja, Kirsten Isensee, Francesca Santoro, Claire Poyser and Wendy Watson-Wright at IOC-UNESCO; and Harriet Thew at WAGGGS for their excellent substantive inputs to the document. We would also like to thank the young ocean activists, Sena Blankson, Luisa Sette Camara, Miriam Justo, Emilie Novaczek, Liam O'Doherty and Marlon Williams for sharing their inspiring stories with us. Special thanks are also due to our wonderfully creative and ever-patient graphic artists, Pietro Bartoleschi (original design), Elisabetta Cremona, Arianna Guida and Fabrizio Puzzilli (design and layout), as well as Simone D'Ercole (photo editing). All of the contributors care deeply about the fate of the ocean and support ocean conservation initiatives. Many thanks also go to the YUNGA ambassadors for their passion and energy in promoting this Guide.

TABLE OF CONTENTS

PREFACE

THE OCEAN IS OUR LUNGS AND LIFE SOURCE.

Welcome to the Blue Planet

The ocean has it all: from microscopic life to the largest animal that has ever lived on Earth, from the colourless to the iridescent, from the frozen to the boiling and from the sunlit to the mysterious dark of the deepest parts of the planet. The ocean is the largest ecosystem on Earth and provides 99 percent of the living space for life. It is a fascinating, but often little explored place.

The ocean affects us in many different ways. It provides us with an important source of food and other natural resources. It influences our climate and weather, provides us with space for recreation and gives us inspiration for stories, art work and music. The list of benefits we get from the ocean is almost endless! But we are also affecting the ocean. Overfishing is reducing fish populations, threatening the supply of nutritious food and changing marine food webs. Our waste is found in massive floating garbage patches and plastics have been found from the Arctic to the bottom of the deepest places in the ocean. Climate change and its related impacts, such as ocean acidification, are affecting the survival of some marine species. Coastal development is destroying and degrading important marine habitats. Even recreation is known to impact marine habitats and species.

We need a clean and healthy ocean to support our own health and survival, even if we don't live anywhere near it. Governments and international organizations are taking action, creating more Marine Protected Areas, improving fishing and environmental regulations and changing the ways that they plan activities in the marine environment. Each and every one of us can also make a difference though. Here are just a few ideas: we can buy only sustainably harvested fish, we can reduce our use of plastics, or we can take part in beach cleans. We can learn more, spread the word and get involved.

This **Youth Guide to the Ocean** gives you a starting point. It illustrates many of the wonders of the ocean, what the ocean does for us and what we are doing to the ocean. It also describes what is being done at the national and international level to protect the ocean. Each chapter contains links and resources for you to explore and learn more. At the end of the Guide, Section D is designed to help you take action. It highlights some of the steps that you can take to develop a successful marine project and invites you to be inspired by the stories of young people around the world who are already taking action.

ÁRNI MATHIESEN
Assistant Director General of Fisheries and Aquaculture Department, FAO

STEPHEN DE MORA
Chief Executive, Plymouth Marine Laboratory

BRAULIO FERREIRA DE SOUZA DIAS
Executive Secretary, CBD

We hope that this Youth Guide to the Ocean will help you to learn about the value of our blue world - and especially to understand that hundreds of millions of people, many of them poor, live in coastal communities and depend on the ocean for food, jobs, their prosperity and their well-being. This Guide will show you how we can support the ocean in return for supporting us. Dive in, enjoy and make a difference!

The ocean is a challenging and intriguing environment to comprehend. A common marine heritage unites the world in history, trade and climate. We all have a stake – the marine resources belong to the global community, both of the present and the future. Thus, understanding the ocean comprises a responsibility for all of us. This Guide serves as an introduction to bring knowledge, care and passion for the ocean and all it contains to the youth of today – our stewards of tomorrow.

Just as the Youth Guide to Biodiversity invites young people to learn and take action to save biodiversity, the Youth Guide to the Ocean encourages you to set your own goals and use your own skills and talents to lead an action project that will help to ensure that the ocean, and the fabulous biodiversity it contains, is protected for future generations.

CBD & YUNGA
AMBASSADORS

ANGGUN
YUNGA AMBASSADOR

"We need to find a way to balance our wellbeing with nature's wellbeing. For instance, eating fish is very good for us – but eating too much fish is not very good for the big blue sea! The ocean acts a highway for enormous loads of important goods – but our shipping industry can have devastating effects on marine life. How can we benefit from the ocean without damaging it? This Guide will help you think about how we can find the best solutions for the complicated issues facing the ocean."

CARL LEWIS
YUNGA AMBASSADOR

"Seas, rivers, lakes, soils, plants, animals, humans; we all rely on each other. It's up to us humans to work out how to live in better harmony with the natural systems that sustain us – and it's up to you to help make that change! While you explore this Guide, think about how the natural ecosystems you are reading about affect your life, and how your life impacts them. What can you do to make this relationship more sustainable? Lead the way!"

DEBI NOVA
YUNGA AMBASSADOR

"Do you enjoy spending time by or in the sea? I do! But the sea does much more for us than just make us feel sunny and happy – it gives us food and oxygen and provides a home for countless unbelievably fascinating living creatures. By convincing your friends and families to help protect beaches, coasts and the ocean, you are helping to protect life on Earth itself! Now that's a cause worth fighting for."

EDWARD NORTON
CBD AMBASSADOR

"Life on earth could not exist without our ocean. Nature has created a matchless balance of marine and terrestrial ecosystems, rich in biodiversity. Besides the sustenance supplied by the world's fisheries, the environmental services provided by ocean coasts alone—including tourism and storm protection—have been evaluated to reach nearly US$26 billion a year. And yet, too little care has been taken to protect this precious resource. Many marine species have been driven to extinction and others are under severe threat, as their habitats are destroyed by human activity and mismanagement. We can and must reverse this trend of destruction, which will impact humanity and our planet for millennia. How we rise to the challenge will define our generation for decades to come."

© FAO/Simone Casetta

FANNY LU
YUNGA AMBASSADOR

"The ocean faces threats from all sides: overfishing, pollution, climate change... And, unfortunately, I'm sure you can think of many more. The important thing is not to feel discouraged: if we put our minds to it, we can turn this around! Step by step, we need to find and promote solutions. Use this Guide to build your understanding of the problems the ocean faces and to decide what part you would like to play in saving the seas."

LEA SALONGA
YUNGA AMBASSADOR

"The ocean is stunning and fascinating. Read this Guide and marvel at the incredible (and sometimes crazy!) adaptations that sea creatures have developed over the millennia to survive in unusual or difficult conditions. Let's make sure our actions on land don't irrevocably harm this beautiful but delicate underwater universe."

© FAO/Simone Casetta

NADÉAH
YUNGA AMBASSADOR

"Little actions go a long way towards conserving the marine environment! For example, you can use fewer plastic bags, try to eat only eating sustainably sourced fish, or make sure to buy coral-friendly sunscreen. How many other ideas can you come up with? Put them into practice and convince as many people as you can to join you – before you know it you'll have started a movement!"

PERCANCE
YUNGA AMBASSADORS

"The ocean connects us to each other: for thousands of years, people have explored the world by navigating the seas. But did you know the ocean connects us digitally too? Most of the internet cables that link us internationally run along the ocean floor! Bear this in mind as you read about the many, interrelated problems that threaten the ocean: together, we are strongest. Raise your voice for the ocean, and see how many others you can convince to join your call to action!"

VALENTINA VEZZALI
YUNGA AMBASSADOR

"Did you know that more of our planet is covered by seawater than land? That's a little reminder for us not to neglect the waters that surround us... The ocean is crucial in the water cycle, nutrient recycling, oxygen production, temperature regulation and many, many other life-supporting services. Life started in the water – let's respect that and nurture our world's waters."

© FAO/Simone Casetta

Download this Guide and other interesting resources from:
www.yunga-un.org

HOW TO USE THIS GUIDE

Throughout this Guide, you will come across little icons. These are a quick way to see what you are reading about:

DID YOU KNOW?

Our world is full of strange and wonderful things. Learn more about it with these fun facts!

? DID YOU KNOW?

BOXES

FIND OUT MORE!

The information contained in these boxes will help you to reflect on issues affecting the ocean.

 FIND OUT MORE!

THEMES

PEOPLE AND OCEAN

Humans both rely on and affect the ocean in many, many ways - these boxes explore the relationship between people and the ocean.

MANAGING THE OCEAN

The ocean is an incredible resource. How can we manage it in a way that benefits both us and the creatures living in it?

STUDYING THE OCEAN

Find out what scientists and researchers are learning from the ocean and what tools and techniques they are using to study it.

Finally, when you see text highlighted like this, you know that the word is in the GLOSSARY at the back of the Guide, where you can look it up for more information.

Section

A

INTRODUCING the OCEAN

Chapter 1

WHAT IS THE OCEAN?

Chapter 2

WHAT DOES THE OCEAN DO FOR US?

Chapter 3

THE OCEAN IN THE PAST

WHAT IS THE OCEAN ?

EARTH IS COVERED BY JUST OVER 350 MILLION KM² OF SALT WATER, EXTENDING OVER ALMOST 72 PERCENT OF THE EARTH'S SURFACE, GIVING IT ITS NICKNAME: THE BLUE PLANET.

1

Kelly-Marie Davidson, Plymouth Marine Laboratory

The ocean is the most essential building block for life. Without the ocean, the Earth would not be habitable for humans, animals or plants. In fact, without the ocean, life would never have begun all those 3.5 billion years ago! Let's take a look at the ocean and its different zones, and answer some key questions about this life-giving feature of the natural world.

EARTH VIEWED FROM SPACE.
© NASA

3

FIVE 'OCEANS'

The global ocean is made up of five interconnecting ocean areas, which are often referred to as individual oceans:

1 ATLANTIC OCEAN: The second largest ocean and on average it is also the saltiest major ocean. It receives <u>fresh water</u> from over half the world's land surface. It is home to the world's first successful under sea telegraph cable, which reduced the communication time between North America and Europe from ten days (the time for a ship to sail across the ocean) to just a few minutes.

2 PACIFIC OCEAN: The largest of the oceans, with an area bigger than all the land masses on Earth combined. It contains the deepest point in the world, the <u>Marianas Trench</u> (found in the Western Pacific) which is just over 11 km deep.

③ Arctic
OCEAN

② Pacific
OCEAN

④ Indian
OCEAN

⑤ Southern
OCEAN

③ ARCTIC OCEAN: The smallest, shallowest ocean. Until recent decades, the Arctic Ocean has almost completely frozen over in the period between October and June. However, <u>climate</u> scientists now predict that Arctic <u>sea ice</u> could soon disappear during the summer due to <u>climate change</u>, and also decrease significantly during the winter. The size of the icesheet varies by about 7 000 000 km^2 between winter and summer.

④ INDIAN OCEAN: The third largest ocean. An estimated 40 percent of the world's offshore oil production comes from this ocean, and delivery of this oil to other parts of the world has been helped by the construction of the Suez Canal, which links the Indian Ocean to the Mediterranean Sea.

⑤ SOUTHERN OCEAN: The fourth largest ocean, which is also referred to as the Antarctic Ocean, South Polar Ocean or Austral Ocean. There is a lot of disagreement about the boundaries of this ocean, as many still consider it to be part of the Pacific, Atlantic or Indian Oceans.

Smaller regions of the ocean, often enclosed by land on more than one side, are known as <u>seas</u>, <u>straits</u>, <u>basins</u>, <u>bays</u> and <u>gulfs</u>. There are more than 100 of these smaller areas, such as the Mediterranean Sea, the Black Sea and the Yellow Sea, as well as a number of <u>inland seas</u> that have salty water and similar characteristics to the five oceanic areas, such as the Aral Sea and the Caspian Sea.

DID YOU KNOW?

:: At least 50 percent of the oxygen in the atmosphere has come from the ocean. This means every second breath you take comes from the ocean.

:: The ocean contains about 97 percent of all the water on the Earth and it is from the ocean that we get our rain water and ultimately our drinking water.

:: The ocean is home to 80 percent of the Earth's organisms; from enormous whales and dolphins to microscopic plankton and bacteria. But scientists believe there are still millions of marine species yet to be discovered!

:: Its average depth is about 4 km.

:: Its average temperature is 2°C and it freezes at -1.8°C.

:: The ocean contains nearly 20 million tons of gold!

:: At the ocean's deepest point, the water pressure is over 10 tonnes per square metre – that's the equivalent of a person trying to carry 50 jumbo jets!!!

:: The longest continuous mountain chain in the known universe, the Mid-Ocean Ridge, is underwater, and stretches for more than 64 000 km.

:: There are over 5 000 active underwater volcanoes.

:: The Great Barrier Reef, measuring over 2 000 km, is the largest living structure on Earth and can be seen from the Moon.

GREAT BARRIER REEF, AUSTRALIA.
© NASA

A WHALE'S TAIL.
© Plymouth Marine Laboratory

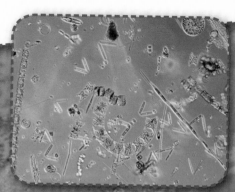

MICROSCOPIC PHYTOPLANKTON.
© Claire Widdicombe, Plymouth Marine Laboratory

WHY THE OCEAN LOOKS BLUE

The colours that make up light scatter (separate and spread) when passing through air or water. Blue scatters more easily than the other colours, so blue rays get reflected back from the surface of the ocean and that's what reaches our eyes. Tiny particles of matter floating in the ocean contribute to the ocean's colour, too. For example, floating plants (phytoplankton) and animals (zooplankton) can give the ocean a more greenish hue, while waters that are a deep, clear blue contain less of this microscopic marine life.

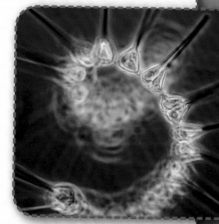

PHYTOPLANKTON.
© NOAA

Earth's atmosphere

Light rays from the Sun

Blue light is scattered

Ocean

SOME SALTY FACTS...

All water contains salts. Different concentrations of these salts make sea water taste salty (saline), while fresh water is much less saline.

:: Of all the water on Earth, only about 4 percent is fresh water.

:: Sea water is approximately 220 times saltier than fresh water.

:: If you could remove all the salt from the ocean (all 46 million km³ of it!) and spread it evenly over the Earth's land, it would reach to the height of a 40 storey building!

:: Most of the salt in the ocean comes from the long-term weathering and erosion of rocks and mountains by rain and streams, but some salt also comes from sediments below the sea floor and material escaping from the Earth's crust through volcanic vents.

SALT PILES IN THE MEDITERRANEAN.
© Svetlana Guineva, Flickr

DRINKING
SEAWATER

Humans cannot drink seawater or use it to water crops without extremely expensive treatments to remove the salt (known as desalination). Marine mammals (such as whales and dolphins) have had to adapt to their salty environment. They generally do not drink seawater and get most of their water through their food, but unsurprisingly, salt does get into their system. These mammals are able to process and release excess salt from their bodies, often by having very salty urine!

Humans haven't adapted in this way: if you drink too much salt water, you will need to urinate more water than you drank to get rid of the excess salt. This actually makes you thirstier and can eventually cause dehydration...

ILLUSTRATION FOR "THE RIME OF THE ANCIENT MARINER" BY GUSTAVE DORÉ.

RIDDLES IN THE WATER

"Water, water everywhere, nor any drop to drink." What do you think the poet Samuel Taylor Coleridge is referring to in his poem *The Rime of the Ancient Mariner* (1798)?

OCEAN ZONES

The ocean is a very dynamic and complex environment. Looking down through the water column, we can see three main layers, or 'zones', that differ according to the amount of light that reaches them:

PHOTOSYNTHESIS

SUNLIT ZONE

200 m

TWILIGHT ZONE

1 000 m

MIDNIGHT ZONE

4 000 m

LOWER MIDNIGHT ZONE

WATER COLUMN.
© YUNGA, Emily Donegan

SUNLIT ZONE

This is the layer at the surface of the sea, up to 200 m deep, where there is enough light for certain <u>organisms</u> to convert the Sun's energy into food, a process known as <u>photosynthesis</u>. <u>Organisms</u> that <u>photosynthesise</u> include microscopic <u>phytoplankton</u>, giant seaweeds and kelps. They are extremely important in marine <u>food webs</u> because they provide food for other creatures such as <u>zooplankton</u>, small fish, jellyfish and some whales. These in turn are eaten by other marine animals like sharks, dolphins and tuna. The sunlit zone supports a mass of marine life.

SNAKELOCKS SHRIMP.
© Matt Doggett, Earth in Focus

TWILIGHT ZONE

From approximately 200 m to 1 000 m, this layer receives less light than the sunlit zone and therefore <u>photosynthesis</u> starts to become difficult. At about 500 m, the amount of oxygen in the water is also greatly reduced (although this varies and in some regions, such as the Arabian Sea and the eastern Pacific Ocean off Peru, there is hardly any oxygen in the water between 100 m and 1 000 m). To survive here, animals have to adapt more efficient ways of breathing or reduce their movement to save energy. Some creatures rise to the sunlit zone at night to feed, avoiding predators and the damaging rays of the Sun. Animals living within this layer include swordfish and <u>bioluminescent</u> (glowing) jellyfish.

BIOLUMINESCENT JELLYFISH.
© Sierra Blakely, WMC

MIDNIGHT ZONE

From 1 000 m and below, the ocean is pitch black apart from occasional light-producing <u>organisms</u>, such as lanternfish. There is no living plant life here. Most animals surviving in this zone rely on '<u>marine snow</u>' for food, a mixture of waste matter and dead <u>organisms</u> falling from the layers above. The giant squid lives in this layer, as do its hunters: deep-diving sperm whales.

LOWER MIDNIGHT ZONE

The lower midnight zone starts at 4 000 m and extends down to just above the sea floor. Very few creatures have successfully adapted to survive the cold temperatures, high water pressure and complete darkness of this zone. Those that have are often very strange – many are transparent and eyeless as a result of the complete darkness! Animals at this depth include some <u>species</u> of squid, and <u>echinoderms</u> including basket stars, swimming sea cucumbers and the sea pig.

FLASHLIGHT FISH.
© Una Smith, NOAA, WMC

ZONES WITHIN ZONES

Each ocean zone contains many different habitats which support a wide range of plants and animals. A habitat is the local environment in which an organism is usually found.

The sunlit zone is home to many well-known habitats such as coral reefs, mangroves, estuaries and rocky shores.

People used to think that very little life survived below the sunlit zone and that the seabed was a flat and featureless place. We now know that there are in fact some very interesting habitats to be found there including deep sea trenches, seamounts, hydrothermal vents and gas seeps.

EUROPEAN LOBSTER HIDING ON THE SEABED.
© Matt Doggett, Earth in Focus

Read on to find out more about them...

CONCLUSION

No wonder we call Earth the 'Blue Planet' – the ocean covers almost 72 percent of its surface! It also supports the vast majority of life on Earth – 80 percent of it, in fact! This includes many weird and wonderful creatures (take a look at the 'Aquarium in an Annex' at the end of the book for a selection of particularly fascinating underwater animals) - and it also includes us.

With the help of this Guide you will learn just how much people rely on the ocean. You will get to see how the different parts of the ocean and the life within it are interconnected, creating a complex ecosystem. You will discover that this ecosystem is fragile, and that our precious ocean faces daunting threats, many of them caused by humans. And you will learn much more about these threats, and about the importance of protecting the ocean for a sustainable future.

LEARN MORE

:: National Geographic Education:
www.education.nationalgeographic.com/education/encyclopedia/ocean/?ar_a=1

:: The National Ocean and Atmospheric Association:
www.noaa.gov/ocean.html

:: National Geographic: www.ocean.nationalgeographic.com/ocean

:: Smithsonian Ocean Portal: www.ocean.si.edu

WHAT DOES THE OCEAN DO FOR US?

THE OCEAN IS OUR LIFE SOURCE!

Caroline Hattam, Plymouth Marine Laboratory

We have already mentioned that, without the ocean, life could not exist. Humans have an even closer relationship with the ocean than we often realize. This chapter introduces the different ways that people use the ocean, and explains why protecting the ocean is a responsibility that each and every one of us needs to share.

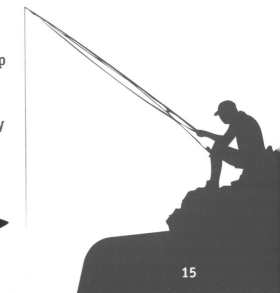

WORKING TOGETHER TO HAUL IN THE FISHING NET.
© Luca Coccia, iStockphoto/Thinkstock

HOW DO WE <u>USE</u> THE <u>OCEAN</u>?

PEOPLE & OCEAN

We use all parts of the ocean: the edge, the surface, the seabed and everything in the middle.

OFF SHORE WIND TURBINES.
© Steve Fareham, WMC

FISHING AND FISH FARMING

CRAB BOAT FROM THE NORTH FRISIAN ISLANDS WORKING IN THE NORTH SEA.
© Jom, WMC

About 56 million people are currently employed in fisheries and <u>aquaculture</u> (fish farming). In addition, many more are employed in follow-up activities, such as handling, processing and distribution. Altogether, fishing and fish farming support the livelihoods and families of some 660 to 880 million people – that's 12 percent of the world's population. More than 40 percent of fish and shellfish eaten by humans come from <u>aquaculture</u>, much of which is found in <u>coastal zones</u> (where it's known as <u>mariculture</u>).

NATURAL RESOURCES

GERMANY'S LARGEST OIL FIELD MITTELPLATE.
© Ralf Roletschek, WMC

Offshore oil and gas rigs currently provide 30 percent of the world's oil production and 50 percent of its gas production.

Marine sand and gravel are mined for use in the construction industry and interest is growing in mining the seabed for metals such as iron, copper, zinc, gold and silver.

RENEWABLE ENERGY

Devices are being developed to generate electricity from waves and <u>tides</u>. Offshore wind farms are also under development, as are <u>biodiesels</u> made from marine <u>algae</u>.

A PLACE TO LIVE

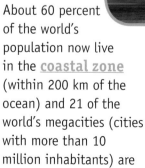

THE GOLD COAST SKYLINE, AUSTRALIA.
© Mike:R, WMC

About 60 percent of the world's population now live in the <u>coastal zone</u> (within 200 km of the ocean) and 21 of the world's megacities (cities with more than 10 million inhabitants) are in coastal areas.

DIVER CHECKING
UNDERWATER
PROTECTION OF CABLE.
© CTBTO Preparatory
Commission

A SURFER IN
SANTA CRUZ.
© Robert Scoble, WMC

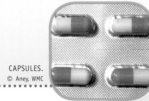

CAPSULES.
© Aney, WMC

COMMUNICATION

Submarine cables are essential for international communication. 99 percent of internet traffic flowing between countries goes via submarine cables!

RECREATION AND LEISURE

It has been estimated that worldwide 121 million people take part in marine recreational activities such as fishing, whale watching and diving each year. This industry is thought to be worth more than US$ 47 billion each year.

MEDICINES

Scientists have discovered that many marine invertebrates produce antibiotic, anti-cancer and anti-inflammatory substances. Horseshoe crabs, seaweeds and marine bacteria have also been found to have useful medical properties.

TRANSPORT AND COMMERCE

90 percent of world trade is carried out by sea! Passenger ferries are also popular modes of transport. In the UK alone, 21.1 million people travelled to and from the UK by sea in 2011.

ORNAMENTAL RESOURCES

Global trade in marine ornamental resources (such as aquarium fish, corals and shells) is estimated to be worth between US$ 200 and 330 million each year. Up to 2 million people worldwide are thought to own marine aquaria.

A SMALL AMATEUR
AQUARIUM.
© Aleš Tošovský, WMC

CONTAINER SHIPS,
SAN FRANCISCO.
© NOAA, WMC

HEALTHY OCEAN, HEALTHY FOOD
Jogeir Toppe, FAO

SUSHI.
© cyclonebill, WMC

Today, your average person eats 19 kg of fish each year (*Source*: FAO, 2012). Foods from the ocean are very important for global food and nutrition security, because eating fish and other seafood has unique nutritional and health benefits. Fish is a key element in a healthy diet as it provides us with good supplies of high quality protein – but it's not the only important **nutrient** the ocean gives us! Most seafood is full of healthy micronutrients (like minerals and vitamins) as well as healthy fats (you might already have heard about omega-3 fatty **acids**). Apart from fish, crustaceans, **bivalves** and plants like seaweed and kelp provide us with these **nutrients** and micronutrients too.

PROTEIN

Why is protein so important? Here we go: every cell in our body contains protein. Protein helps build and repair our body's tissues (muscles, bones, organs, skin and hair) and is needed for clotting blood, fighting diseases and for producing hormones, among many other things. It's particularly important for healthy growth and development during

↗

childhood, adolescence and pregnancy. Are you convinced now?

About 17 percent of the animal protein the global population eats comes from fish. However, in some countries in West Africa or Asia, for example, this share is much higher (often 50 percent and above), making fish a crucial part of many local diets.

FATS

Did you know that your brain is basically made of fish oil? Well, that's a slight exaggeration, but the essential fatty **acids** found in seafood (known as 'long chain omega-3 fats') are essential for the healthy development of the human brain, and are therefore especially important for babies and growing children. That's also why it's also important for pregnant women or mothers who are breastfeeding to eat a minimum amount of fish: it enriches the food their children get from them.

Many vegetable oils also contain omega-3 fatty **acids**, but it's a different kind of omega-3 (called alpha-linolenic **acid**) which is short in length and needs to be converted into a longer form before our brains can really use them. This makes them more difficult for the brain to rely on than fish-based omega-3.

Omega-3 fatty **acids** from seafood are also good for adult brains: scientists have found that mental disorders, such as depression and dementia, are less likely in people who eat enough seafood.

Apart from our brains, our hearts also benefit from fatty **acids**: they can help reduce the risk of fatal coronary heart diseases by up to 36 percent.

Ideally, a child's diet should include 150 mg (milligrams) of long-chain omega-3 fatty **acids** a day, and an adult's should contain 250 mg.

MICRONUTRIENTS

Food from the ocean is also a great source of essential micronutrients such as vitamins and minerals. This is in particular true for small-sized seafood that can be eaten whole (heads, bones and everything!), because these can be an excellent source of vitamins (such as A and D) and minerals (including iodine, selenium, zinc, iron, calcium, phosphorus and potassium). Interestingly, different **species** of fish and even different parts of the same fish can contain very different types and amounts of micronutrients.

Micronutrient deficiencies make hundreds of millions of people around the world ill, in particular women and children in **developing countries**. For instance, the latest World Health Organization (WHO) statistics tell us that:

:: More than 250 million children worldwide are at risk of vitamin A deficiency, which can cause blindness or be fatal;

:: The mental health of 20 million people is affected as a result of iodine deficiency (like long-chain omega-3 fatty **acids**, iodine is essential for healthy brain development);

:: 2 billion people (over 30 percent of the world's population) are iron deficient, which can make them feel chronically weak and dizzy and, in severe cases, lead to death;

:: 800 000 child deaths per year can be attributed to zinc deficiency.

Rural diets in many countries may not be particularly varied (meaning people eat a lot of the same things, because there aren't so many options). That means that it is vital for them to have good food sources that can provide all essential **nutrients** they need – and seafood is an ideal provider of many of these.

EATING HABITS

We have seen that replacing foods with lower nutritional values with fish would be a great way to make people healthier, and provide nutritious food to those who are currently going hungry. At the same time, we need to make sure the marine environment stays healthy too. Unfortunately, the demand for fish over the last couple of centuries has grown so rapidly, that the populations of popular fish (like tuna, cod or salmon) are unable to keep up and are decreasing. Sustainable mariculture (fish farming in the sea) is part of the solution to avoiding overfishing wild ocean fish stocks (find out more about mariculture in Chapter 4, and more about overfishing in Chapter 10). Another part of the solution lies with the consumer: each and every one of us can help marine ecosystems by not buying seafood that is under pressure, and choosing sustainably sourced seafood instead. You can make a start by taking a look at National Geographic's Seafood Decision Guide to help you choose which fish to buy or to avoid: **www.ocean. nationalgeographic. com/ocean/take-action/ seafood-decision-guide**

Find out more:
:: FAO Fisheries:
www.fao.org/fishery/en
:: Feeding Minds, Fighting Hunger: Fisheries and Aquaculture:
www.feedingminds. org/fmfh/fisheries- aquaculture/wonders- of-the-oceans/en

FISH FOR SALE AT A FISHMARKET.
© Kelvin Boot

WHAT ELSE DOES
THE OCEAN DO FOR US?

PEOPLE & OCEAN

As well as giving us food, raw materials, jobs, transport, energy and opportunities for leisure, the ocean does many more things for us that we often take for granted. These benefits that the ocean provides for humans are known as ecosystem services. For example:

THE OCEAN REGULATES OUR CLIMATE: We've already said that we get oxygen from the ocean, but did you know that the ocean has also absorbed a quarter of all the carbon dioxide that humans have put into the atmosphere? This makes the ocean a 'carbon sink' (somewhere carbon dioxide is stored). Without the heating and cooling effects of ocean currents, world temperatures would be too unstable to support life.

IT AFFECTS OUR weather: As the ocean is heated by the Sun's rays, water from its surface evaporates and then condenses to form clouds as part of the water cycle. This is how we get our rain and therefore our drinking water. It also contributes to wind, thunderstorms and hurricanes, and helps produce the monsoon rains upon which millions of people in South Asia rely!

IT TAKES CARE OF MANY OF OUR WASTE PRODUCTS: The ocean dilutes and disperses our waste products, while marine animals can bury them, absorb them within their own bodies or break them down into harmless substances. This may be good for us, but can do great damage to ocean life: unfortunately, humans have increasingly treated the ocean as a dumping ground. You can find out more about the ocean and pollution later in the Guide.

KIDS ON THE BEACH.
© voldevis, www.stockvault.net

TORNADO OUT AT SEA.
© NOAA

IT INFLUENCES OUR HEALTH AND WELL-BEING: Have you noticed an aquarium in your doctor's or dentist's surgery? Water is known to calm and reduce anxiety in people and being near blue spaces, such as the ocean, is thought to have positive effects on our mental health.

IT IS A SOURCE OF CULTURAL INSPIRATION: The ocean and the <u>biodiversity</u> within it is a source of inspiration for art, poetry, novels, songs and folklore. Can you think of an ocean-inspired story, song or piece of artwork?

THE GREAT WAVE OFF KANAGAWA.
Katsushika Hokusai (1829-32)

MORAY BURROWING ON THE SEABED.
© Efraimstochter, www.all-free-download.com

WHY DO WE NEED TO
PROTECT
THE OCEAN?

MANAGING THE OCEAN

:::: Different human
activities are putting
the ocean under threat: ::::

COASTAL DEVELOPMENT: The building of houses, hotels, roads and industrial sites is responsible for the degradation of many coastal marine ecosystems such as coral reefs, seagrass meadows and mangroves, yet it is these ecosystems that help to protect vulnerable coastlines and the people living there against storm surges.

POLLUTION: Approximately 80 percent of the pollution in the ocean comes from land, and coastal zones are especially vulnerable to pollutants. Plastics are also particularly problematic with enormous floating rubbish patches forming in the ocean. The Great Pacific Garbage Patch, for example, is a huge floating mass of rubbish stuck in a gyre (a circular ocean current) in the North Pacific Ocean.

DID YOU KNOW?

We dump about three times more rubbish into the ocean every year than the weight of the fish we catch annually... And each year more oil enters the ocean as runoff from roads than from accidental oil spills.

THIS PAGE
PLASTIC BAG ON THE SEABED.
© Matt Doggett, Earth in Focus

FACING PAGE
LIONFISH, AN INVASIVE SPECIES.
© Alexander Vasenin, WMC

FISH FARM IN BOLIVIA.
© Christopher Walker, WMC

FISHING: 70 percent of the world's fish <u>species</u> are either already <u>overfished</u> (too many are caught) or would become <u>overfished</u> if catch rates increase. Some fishing activities are damaging fragile marine <u>ecosystems</u>, such as bottom <u>trawling</u>, which destroys seabed communities.

AQUACULTURE: The global increase in <u>aquaculture</u> (fish-farming) while being an important source of food, has in some places led to coastal <u>habitat</u> degradation and loss, pollution, the introduction of exotic <u>species</u> and the spread of diseases among humans and animals.

INVASIVE SPECIES: Marine <u>species</u> are transported around the world as secret 'stowaways' on <u>cargo</u> ships, attached to ships hulls or contained in their <u>ballast</u> <u>water</u>. When they enter new environments they can cause problems for native <u>species</u>, as they may be better at finding food and shelter.

CLIMATE CHANGE, OCEAN ACIDIFICATION AND <u>HYPOXIA</u>: Changing ocean temperature, <u>acidity</u> and oxygen levels are affecting the distribution of marine <u>species</u> and their ability to grow and reproduce. This is a particularly serious concern for the ocean's magnificent but fragile coral <u>reefs</u>, which support vast amounts of marine life.

MARINE DEBRIS FROM BELOW.
© NOAA

WHO IS RESPONSIBLE AND WHAT IS BEING DONE?

Individual countries are responsible for protecting the ocean in their own territorial waters and exclusive economic zones. They can regulate their fisheries (often together with other countries), designate Marine Protected Areas and attempt to control coastal development and pollution.

In the high seas, where no particular country is 'in charge', many different international bodies set the rules. For example, the International Seabed Authority is responsible for developing deep sea mineral resources and protecting the seabed from mining activities; the International Maritime Organization is trying to reduce the spread of invasive species and control pollution; and Regional Fisheries Management Organizations set fishing regulations. As the global ocean is interconnected, the Intergovernmental Oceanographic Commission promotes international cooperation and coordination in research, services and capacity building to improve ocean and coastal knowledge and management.

In addition to these national and international activities, each and every one of us has a role to play in protecting the ocean. We can buy sustainably sourced fish; reduce our use of plastics, such as bags and bottles, that may end at sea; buy cleaning products that are safe for marine and other aquatic organisms; and we can take part in beach cleans. There are many, many ways to get involved and you can find more ideas in Chapter 14.

OCEAN TERRITORIES

Defined by the United Nations Convention on the Law of the Sea (UNCLOS), which has been in force since 1994:

:: Territorial waters *extend 12 nautical miles (22 km) from the coast;*

:: Exclusive economic zones *extend up to 200 nautical miles (370 km) from the coast.*

CONCLUSION

So, we can see that without the ocean, we wouldn't be here! Not only does the ocean provide us with food, water, medicines and other resources, but it also provides many other essential 'ecosystem services': the ocean makes sure that our climate is neither too hot nor too cold, it manages some forms of human waste and it provides inspiration for art, music and poetry.

But humans are putting too much pressure on the ocean. Climate change and pollution are starving the ocean of oxygen, changing its temperature and levels of acidity, and threatening its precious biodiversity. Overfishing has put many species of fish on the endangered list. And human activity at the coast is changing coastal habitats. If we carry on mistreating the ocean, we will lose its vital services. But there are ways you can help. Read on for more inspiration...

LEARN MORE

:: Intergovernmental Oceanographic Commission of UNESCO: **www.unesco.org/new/ioc**

:: International Maritime Organization: **www.imo.org**

:: International Seabed Authority: **www.isa.org.jm**

:: Marine Conservation: **www.marine-conservation.org**

:: Marine Conservation Society: **www.mcsuk.org**

:: National Geographic on protecting the ocean: **www.ocean.nationalgeographic.com/ocean**

:: Regional Fisheries Management Organizations: **www.ec.europa.eu/fisheries/cfp/international/rfmo**

:: United Nations Convention on the Law of the Sea:
   www.un.org/depts/los/convention_agreements/texts/unclos/UNCLOS-TOC.htm

:: World Wildlife Fund: **www.wwf.panda.org/about_our_earth/blue_planet/open_ocean/ocean_importance**

THE OCEAN IN THE PAST

THE OCEAN HAS SEEN MANY CHANGES THROUGH TIME AND HAS SHAPED HUMAN HISTORY.

Kelvin Boot, Plymouth Marine Laboratory

The Earth is unique in being the only planet in our solar system that has large amounts of liquid water on its surface. But where did it come from?

A FOSSIL OF A FISH (AIPICHTHYS).
© WMC

THE SOURCE OF **EARTH's WATER**

STUDYING THE OCEAN

Most scientists agree that water first appeared on Earth as soon as the conditions of temperature and pressure allowed it to exist in a liquid state: this would have been around half a billion years after the Earth was formed, about 4 billion years ago. What scientists are less sure about is where it came from.

There are a number of theories about how the Earth got its water:

A long-held theory is that hydrogen and oxygen atoms combined into water molecules within the Earth as the planet was forming, and this reached the surface when it was expelled along with magma from volcanoes.

#1

Some believe that fully formed water arrived from space on the dust particles from which the Earth itself was formed, again reaching the surface through volcanic eruptions.

#2

BARRINGER METEOR CRATER IN ARIZONA.
© WMC

TWO FRAGMENTS OF THE CHELYABINSK (CHERBAKUL) METEORITE.
© WMC

COMET P1 MCNAUGHT.
© Fir0002-Flagstaffotos, WMC

Alternatively, others have suggested that comet impacts, which were common in the Earth's youth, brought water to its surface. This theory followed the discovery that comets are composed of ice as well as rock.

#3

Asteroids and meteorites are other interplanetary travellers: many of these also carry water, but critics of this theory point out that it would take an awful lot of meteoritic water to fill an ocean! Having said that, Earth was subjected to almost non-stop bombardment for thousands of years while it was forming...

#4

So the jury is still out, and it may turn out that each of these origins might have contributed to Earth's water supplies.

THE <u>EVOLUTION</u> OF THE OCEAN

When we look at a map of the Earth (or, better still, a <u>satellite</u> view across the South Pole), it is obvious that today's ocean is one vast stretch of water.

The ocean has remained more or less the same in size, shape and coverage throughout the time that humans have inhabited this planet. There have been minor variations as <u>glacial periods</u> have come and gone in the recent <u>geological</u> past (recent being the last million years in this case!), but these have had little impact on the overall area covered by the global ocean.

The arrangement of land and water, on the other hand, has changed considerably over the past few billion years. We know very little about the early ocean,

SOUTH POLE VIEWED FROM SPACE.
© NASA

its layout and position on the globe, so much guesswork is involved in trying to draw maps for the period. Sophisticated chemical analysis of extremely old rocks that have survived from long ago provides some evidence of later ocean layouts, but even these are questionable. It is only in the last billion years that the relative positions of ocean and land can be given with any confidence.

PSEUDOASAPHUS
(A TYPE OF TRILOBITE).
© WMC

DID YOU KNOW?

There are many different kinds of evidence proving the theory that Earth's land masses moved apart and reunited in differing arrangements through time. 'Magnetic fingerprints', for example, are patterns that show how the direction of the Earth's magnetic field has changed in the past. Other clues are provided by the presence of identical <u>fossil</u> remains on continents that are now thousands of miles apart. Just consider the shape of today's continents, you also get a sense of how they once fitted together – just take a look at the diagram on pp.34-35!!

VIEW FROM THE ROOI ELS COASTAL ROAD,
SOUTH AFRICA.
© Danie van der Merwe, WMC

SUPERCONTINENTS
AND THE OCEAN

Geologists suggest that throughout geological time there have been a series of 'supercontinents', where instead of there being several continents as there are today, a single giant landmass was surrounded by water. There is still much dispute over how many supercontinents there were and when and where they were found, though.

Perhaps the best known and most widely accepted supercontinent is Pangaea and its surrounding ocean Panthalassa. Pangaea existed between 270 and 200 million years ago, and can be thought of as a giant jigsaw puzzle with today's continents as its pieces.

It is particularly easy to imagine the east coast of South America fitting against

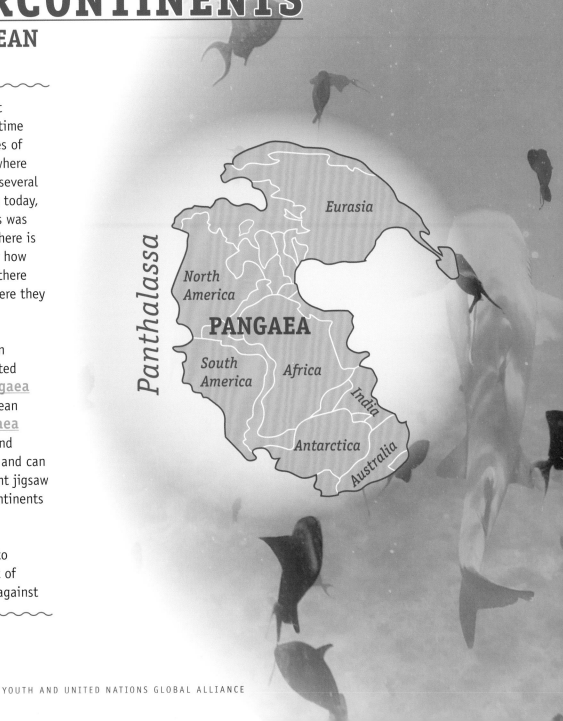

North America

LAURASIA

Eurasia

South America

Africa

Tethys Sea

GONDWANA

India

Antarctica

Australia

Africa's west coast, and it was this observation that lead to early theories that the continents were not fixed in place but moved around the Earth. This process is known as <u>continental drift</u>. <u>Continental drift</u> tore Pangaea apart (very slowly!) to form the smaller supercontinents of Laurasia in the north and Gondwana to the south, with the Tethys Sea, later to become the Tethys Ocean, in between.

The first modern ocean areas formed around 180 million years ago as northwest Africa and North America separated, creating the central Atlantic <u>basin</u>. Some 40 million years later the South Atlantic was added as Africa and South America drew apart. The Indian Ocean was formed at about the same time, when India, Antarctica and Australia parted.

The final pieces of the jigsaw, North America and Europe, moved away from each other about 80 million years ago.

SHARK AND FISH.
© Suzanne Redfern

CHANGING SEAS

The Tethys Ocean spanned the globe throughout the Mesozoic Era of geological time, from about 250 million to 60 million years ago. During this period of global history, the land was populated with gargantuan dragonflies, the dinosaurs, and, towards its end, with the first birds and mammals.

The Tethys Ocean teemed with life, including marine reptiles such as the long-necked plesiosaurs and the ichthyosaurs, which looked a lot like modern dolphins. This was a period of relative calm in the ocean's chemistry and geology, and these stable conditions allowed marine animals to flourish.

Among the molluscs, spiral-shelled ammonites developed into an astonishing variety of species leaving only their shells behind as fossils; clues to their now extinct lives.

A SEASLUG AMONGST ANEMONES AND BRYOZOANS.
© Matt Doggett, Earth in Focus

Corals too flourished in the warm clear waters, leaving evidence of their presence in rocks across the globe – even in areas now close to the poles, providing even more evidence for the movement of the continents.

This period of stability came to an end around 65 million years ago when the Earth's land and ocean experienced one of their greatest changes. The led to a mass extinction, which did not just destroy the dinosaurs, it eradicated up to 70 percent of all known species including: 90 percent of algae species; 98 percent of corals; 35 percent of echinoderms and 50 percent of all seabed living organisms.

Not all types of marine life were affected equally: surprisingly, nine out of ten species of bony fish survived, but scientists don't really know why.

It has been estimated that recovery took at least tens of thousands of years (perhaps even as much as 2 million years!), but despite life regaining its place, the ocean was very different.

Extinction is a crucial part of evolution (there had been four previous mass extinctions in Earth's history) and this event marked the end of one era and the beginning of our modern ocean...

FROM TOP
A YELLOW-MARGINED
MORAY EEL;
SEAFAN WITH
BRITTLESTAR.
© Matt Doggett, Earth in Focus

BACKGROUND IMAGE
GREY SEAL.
© Matt Doggett, Earth in Focus

THE OCEAN IN
HUMAN HISTORY

PEOPLE & OCEAN

People have been around for only a fraction of the life of the ocean, but the ocean has always been important to us: <u>seafaring</u> has long been part of our history.

CHRISTOPHER COLUMBUS
© Metropolitan Museum of Art, WMC

Early humans are thought to have followed the coast as they **migrated** from Africa into Asia and very similar stone cutting tools have been found in northwest Africa and in Spain, suggesting that people may have crossed the Strait of Gibraltar (almost 10 km, even at the lowest sea level) as early as 500 000 years ago. There is also evidence that people have travelled by sea around the eastern Mediterranean for over 100 000 years.

Longer, open ocean crossings were already taking place in southeast Asia 40 000 years ago, and epic voyages, such as between what is now Indonesia and Madagascar (a distance of around 10 000 km) were undertaken more than 1 000 years ago.

Chinese ships were trading with their neighbours at a similar time, and by the fifteenth century were travelling to east Africa and the Arabian Gulf.

The fifteenth century also saw the Age of Discovery in European maritime history, and included Christopher Columbus's voyage to America, as well as trading expeditions across the globe.

An important part of **seafaring** has always been fishing. Fishing from the coast is thought to have started around 140 000 years ago, and evidence of people fishing from boats dates back to about 40 000 years ago. Today, fishing is the most widespread human activity in the ocean, and its scale ranges from the use of dugout canoes and spears in coastal waters to expeditions to the open ocean in sophisticated factory ships that process the fish caught by a fleet of smaller boats.

Today our use of the ocean is extensive – and keeps growing (as we saw in Chapter 2). The trouble is, the many uses we make of the ocean are changing the ocean's characteristics and the plants and animals that live there faster than any of the natural changes that took place slowly, over long periods of time. You can learn more about these changes and what we can do to help protect the ocean and ocean life in the following chapters.

CONCLUSION

The history of the ocean is the history of the Earth. Throughout its existence, the ocean has kept changing, with continents coming and going. Marine life has survived these sometimes catastrophic changes, although what you see in the ocean today is very different to what you would have seen millions of years ago.

Humans are also part of this history and we have relied on the ocean since our species evolved on Earth, using it as a source of food, for trade and transport and exploration. This ever-increasing use is changing the ocean again and this change may not be for the better.

LEARN MORE

:: BBC Nature: www.bbc.co.uk/nature/extinction_events

:: MarineBio Conservation Society: www.marinebio.org/oceans/history

:: National Geographic:
www.channel.nationalgeographic.com/channel/videos/history-of-the-oceans

:: Wiley:
www.wiley.com/college/strahler/0471480533/animations/ch13_animations/animation3.html

Section B

COASTAL ZONES: GATEWAYS TO THE OCEAN

Chapter 4
WHERE THE OCEAN MEETS THE LAND

Chapter 5
COASTAL WATERS AND THE SEABED

Chapter 6
THE HIDDEN WORLD OF ESTUARIES

Chapter 7
LIVING BETWEEN THE TIDE LINES

Chapter 8
MANGROVES AND SALTMARSHES

Chapter 9
CORALS AND SEAGRASSES

WHERE THE OCEAN MEETS THE LAND

THE COAST IS THE MEETING PLACE BETWEEN LAND, HUMANS AND THE OCEAN, A PLACE WHERE NOTHING EVER STAYS THE SAME...

Caroline Hattam, Plymouth Marine Laboratory

4

The result of constant change in the ocean, a continuous supply of building materials from the land and rivers, and ever increasing human activities means the coastal zone is always in motion. This chapter will give you an introduction to this dynamic and diverse environment, what lives there and how we are affecting it.

Coastal areas are often rich in nutrients and support a great diversity of marine life. They are also exceptionally important to humans and have shaped our relationship with the ocean. An exact coastline cannot be identified because the rising and falling of the tides mean that the coast is never the same from one tide to the next. Instead we use the term coastal zone to describe the area where land influences the ocean and the ocean influences the land.

SURFER RIDING A WAVE.
© Brocken Inaglory, WMC

TROPICAL BEACH.
© Matt Doggett, Earth in Focus

DID YOU KNOW?

Altogether there are about 356 000 km of coast around the world – about the same as the distance between the Earth and the Moon!

HOW TIDES WORKS

Tara Hooper, Plymouth Marine Laboratory/Marine Education Trust

Sea level is not constant. As the hours pass, sea level will rise and fall (also known as <u>flooding</u> and <u>ebbing</u>) in a way that is rhythmical and entirely predictable. <u>High tide</u> (or high water) describes the time when the <u>tide</u> has come as far as possible up the beach, while <u>low tide</u> (or low water) is when the largest amount of beach is visible.

This regular change in the height of the <u>tide</u> – the <u>tidal cycle</u> – is caused by <u>gravitational</u> forces generated as the Moon travels around the Earth and the Earth orbits the Sun. The Moon has the largest <u>tide</u>-generating force (more than twice that of the Sun) because, despite being 27 million times smaller than the Sun, it is nearly 400 times closer to the Earth.

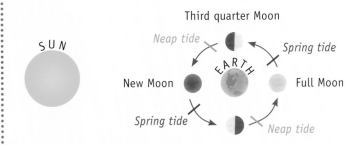

Source: YUNGA

The amount of beach that will be covered and uncovered by the <u>tide</u> is not the same every day. Very <u>high</u> and very <u>low tides</u> (<u>spring tides</u>) happen at full and new Moon, when the <u>gravitational</u> forces of the Sun and Moon combine. <u>Neap tides</u> occur in between <u>spring tides</u>, and the difference between <u>high</u> and <u>low tide</u> is smaller. <u>Neap tides</u> occur when the Sun and Moon are perpendicular: each slightly cancels out the <u>gravitational</u> influence of the other, reducing the difference between the <u>tides</u>. The transition from <u>spring tides</u> to <u>neap tides</u> is gradual, with the <u>tidal range</u> (the difference between <u>high</u> and <u>low tide</u>) becoming a little smaller every day (and getting slowly larger every day as the cycle moves back towards <u>spring tides</u>).

A great deal of energy is contained in the movement of the <u>tides</u>, and people are trying more and more to harness this to generate electricity: through <u>tidal barrages</u>, which exploit the difference in height between <u>high</u> and <u>low tide</u>, or <u>tidal current turbines</u>, which are driven by the speed of the water passing by them.

RICHNESS
AND DIVERSITY

The special relationship between land and the ocean has led to the development of different habitats in coastal zones around the world. These include coral reefs and sea grass meadows, mangroves and saltmarshes, estuaries and deltas, as well as rocky, sandy and shingle shores. The seabed itself is also a very important habitat providing a home to marine life at its surface, and even for several centimetres (and in some cases metres) into the mud or sand as well.

OPPOSITE PAGE
CORAL REEF ECOSYSTEM AT PALMYRA ATOLL.
© Jim Maragos, U.S. Fish and Wildlife Service

TOP IMAGES - FROM LEFT TO RIGHT
INTERTIDAL MUDFLATS EXPOSED AT LOW TIDE.
© Walter Siegmund, WMC

SANDY SHORE ON THE COAST OF ANGOLA.
© Alfred Weidinger, WMC

EXPOSED ROCKY SHORE.
© N. Aditya Madhav, WMC

MAGNIFICENT BRAIDED RIVER DELTA WITH RED AND GREEN FLORA SHOWING AT LOW TIDE. ALASKA, LOWER COOK INLET, KACHEMAK BAY.
© NOAA

MANGROVE FOREST.
© Matt Doggett, Earth in Focus

BOTTOM IMAGES - FROM LEFT TO RIGHT
DIVER IN KELP FOREST.
© Ed Bierman, WMC

SEAGRASS MEADOW.
© Paul Asman, Jill Lenoble, WMC

ROCKPOOL IN THE INTERTIDAL ZONE.
© Brocken Inaglory, WMC

CHUBUT STEAMER DUCK.
© CHUCAO, WMC

HECTOR'S DOLPHIN.
© James Shook, WMC

MACKEREL SCHOOL.
© NOAA

The creatures you find living at the coast will vary depending on the water depth, water and atmospheric temperatures, exposure to waves, water turbulence, sunlight levels, available **nutrients**, **salinity** (saltiness), the presence of dissolved gases and **acidity** levels.

Some marine **species** are said to be **cosmopolitan** and can be found all around the world, such as some **species** of **plankton**. Others are considered **endemic** and are unique to specific **habitats** and locations; for example Hector's dolphins are only found in New Zealand and the Chubut steamer duck, a flightless marine duck, is only found in Patagonia. Many marine **species** are somewhere in between. For example, mackerel are found throughout the ocean in the **northern hemisphere** and the Emperor nautilus can be found from Japan to Fiji to Indonesia and Australia.

SUMMER BLOOM OF MARINE
PHYTOPLANKTON IN THE
BALTIC SEA.
© Richard Petry, Flickr

NAUTILUS IN BERLIN ZOO
AQUARIUM.
© J. Baecker

SEABIRDS

Tara Hooper, Plymouth Marine Laboratory/Marine Education Trust

Seabirds are an important part of our coastal marine life. They prey on crustaceans, mussels, fish, and plankton, so the shore and shallow waters are important feeding grounds. The coastline provides essential breeding habitat too. Noisy colonies of seabirds including gulls, frigatebirds and gannets can be found nesting on cliff ledges, or, in the case of puffins, penguins, storm petrels and auklets, digging burrows in the sand or soil of cliffs, dunes and the top of beaches. Seabirds species such as terns, noddys and albatross also nest in or under shoreline trees.

Some species of seabird undertake epic annual migrations: sooty shearwaters travel over 60 000 km between New Zealand and the North Pacific, and Arctic terns cover a similar distance as they journey between the Arctic and Antarctic. Other birds (e.g. frigatebirds and albatross) do not make set annual migrations, but travel tens of thousands of miles over the ocean as they forage for food.

These long-distance travellers can be affected by human activities occurring in any part of their range, for example, the loss of prey species from their feeding grounds, the destruction of habitat in their breeding areas or disturbance by people carrying out recreational activities. International agreements are therefore required to ensure that seabirds are adequately protected. One area such international co-operation focuses on is reducing the by-catch of seabirds in fishing gear (particularly long lines), through specific agreements such as the FAO Code of Conduct for Responsible Fisheries, National Plans of Action and Regional Fisheries Management Organizations.

PUFFIN SWIMMING UNDERWATER.
© Matt Doggett, Earth in Focus

A FULMAR CHICK IN ITS NEST, SCOTLAND.
© Matt Doggett, Earth in Focus

BACKGROUND IMAGE
SEABIRD COLONY.
© Matt Doggett, Earth in Focus

PEOPLE
AND THE COAST

Most people only get to see the ocean at the coast and it is here that the majority of human activity in the ocean goes on. Many of these uses, such as fishing, leisure and recreation, transport and commerce were introduced in Chapter 2.

People have been using the coast for millennia and this constant use has changed it and the biodiversity found there. In recent decades, as the human population has grown, more and more people are living in coastal zones. As a result, human dependence on marine resources and the coast has increased, and many fragile coastal habitats are under threat or are disappearing as a result of human activities.

AMALFI COAST, ITALY.
© wallpaperpassion

YOUTH AND UNITED NATIONS GLOBAL ALLIANCE

MARICULTURE

José Aguilar-Manjarrez, Alessandro Lovatelli, Doris Soto & Jogeir Toppe, FAO

As the world's population continues to grow, competition for our fresh water resources is increasing. By 2050, it is projected that Earth will need to feed 9.2 billion people. To meet this need for food, we can look to the ocean, but it is unlikely that we can sustainably increase fishing for wild fish (read more about overfishing in Chapter 12). Instead, marine farming, also known as 'mariculture' is a promising option. By expanding the mariculture industry, we could harvest enough healthy food to feed Earth's growing population.

TERMINOLOGY: AQUACULTURE AND MARICULTURE

The terms 'aquaculture' and 'mariculture' both relate to growing or rearing aquatic organisms in a confined environment. You can think of this as similar to how a farmer grows a field of corn or raises a herd of goats. 'Aquaculture' refers to the whole spectrum of fish and shellfish produced in both fresh water and saltwater environments. 'Mariculture' is the branch of aquaculture undertaken in marine (saltwater) environments. The majority of mariculture farms are found in coastal waters, just off the shore.

LIVELIHOODS

Mariculture is already widespread around the world: it is practiced in over 93 countries and territories, farming hundreds of species. Over 23 million tonnes of seafood, excluding marine plants (seaweeds), are produced every year!

FEEDING-TIME IN A GILTHEAD SEABREAM CAGE. IZMIR BAY, TURKEY.
© Ozgur Altan

CAGE MAINTENANCE.
© Oceanspar

Mariculture currently contributes to 17 percent of the total fish that humans eat. There is potential for many more countries to develop their mariculture sectors.

HOW IT WORKS

Mariculture technology has developed considerably over the past 30 years. Fish can now be farmed in floating or submerged net cages placed just off the coast. These cages help to avoid the potentially destructive force of bad weather conditions, and minimize the impacts of the farm on other users of marine space. Some fish farms even use remote underwater cameras and dedicated computer software to allow for automated feeding and fish monitoring, which is both efficient and rather cool!

Let's look at salmon. The Atlantic salmon is the most farmed marine fish (in 2012, 1.4 million tonnes of it was produced). You may be surprised – isn't salmon a fresh water species? Not quite. Salmon are a 'diadromus' species, which means that they reproduce and spend their early lives in a fresh water environment. Then salmon migrate to the sea where they grow to their full size. Commercial salmon farming almost always takes place in marine net cages. Salmon farmers understand the salmon life cycle and can control reproduction to produce millions of young fish in much the same way that farmers reproduce sheep and cattle.

Fish farmers can also artificially reproduce many other species of fish, as well as numerous crustacean species (e.g. shrimp and crabs), molluscs (e.g. oysters, mussels, abalones) and marine seaweeds (e.g. kelp).

CONCERNS AND SOLUTIONS

	Potential impacts or problems of mariculture	What can we do about it?
Health	Does the quality of water, feed or the possible misuse of veterinary drugs make farmed fish a less healthy food than their wild relatives?	Many of the factors that might impact the quality and nutritional value of fish can be monitored and controlled in a farming system, so sustainably farmed fish can be even better for human consumption than wild fish. Of course it is necessary to ensure that international health standards are observed on farms.
Waste	Many fish or shellfish kept in a limited space will produce a lot of waste, which can pollute surrounding fresh water, seawater and groundwater supplies.	• Regularly changing the location of net cages allows marine areas to recover and prevent them from being overexploited. • Choose mixed mariculture, where producers farm fish alongside shellfish (e.g. mussels), who feed by filtering the water around them, removing organic waste. • Farm seaweeds as well because they can absorb excess nutrients produced by fish farms.
Fish food	Bigger, often carnivorous fish (like salmon and tuna) are often fed on sandeels and anchovies, which are wild caught, pressurising wild stocks. In some cases the volume of fish eaten is greater than the volume of fish produced.	Fish by-products and waste, rather than whole fish are being used to make fishmeal and fish oil used to feed carnivorous fish. The amount of feed needed to raise a kilogram of fish tends to be lower than for land animals: so well-managed mariculture can represent a reliable and sustainable source of fish.
Habitats	Shrimp farming in particular has led to the deforestation of mangroves to create space for shrimp ponds (see Chapter 8).	Bans on mangrove-clearing are increasingly being implemented in shrimp farming countries.

FUTURE FISH

How much fish will we be able to farm in the future? This is a difficult question, and its answer depends on many factors. It is possible to estimate the potential for culturing some individual fish species. For example, by identifying marine areas with favourable environmental conditions (e.g. the right water depth, water temperature and current speed) for farming cobia (see map on p.54), and assuming the right technology and easy access from the shore to the offshore farm site (e.g. by boat), you can fairly accurately predict your yields.

There is great potential for expanding farming of bivalves (such as mussels, clams and oysters) and

COBIA.
© Daniel D. Benedetti

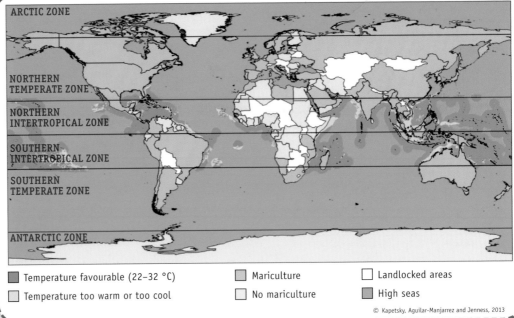

ARCTIC ZONE

NORTHERN TEMPERATE ZONE

NORTHERN INTERTROPICAL ZONE

SOUTHERN INTERTROPICAL ZONE

SOUTHERN TEMPERATE ZONE

ANTARCTIC ZONE

- Temperature favourable (22–32 °C)
- Temperature too warm or too cool
- Mariculture
- No mariculture
- Landlocked areas
- High seas

© Kapetsky, Aguilar-Manjarrez and Jenness, 2013

seaweeds too. And because bivalves are filter feeders, which means they filter seawater for the natural food it contains (such as microalgae), and don't need active feeding, unlike most farmed fish (like salmon), it would be much cheaper! It may appear that there is limitless space for the development of mariculture in our large blue ocean. There are, however, many competing uses of near shore coastal waters. Comprehensive and coordinated area plans need to be put in place to ensure that there is enough space for sustainable mariculture growth and this needs to be planned and managed in harmony with the ecosystem as a whole.

FIND OUT MORE:

:: Building a wooden cage for fish farming: **www.fao.org/docrep/018/i3091e/i3091e.pdf**

:: FAO Fisheries: **www.fao.org/fishery/en**

:: Feeding Minds, Fighting Hunger: Fisheries and Aquaculture:
   www.feedingminds.org/fmfh/fisheries-aquaculture/wonders-of-the-oceans/en

:: The Seafood Decision Guide:
   www.ocean.nationalgeographic.com/ocean/take-action/seafood-decision-guide

CONCLUSION

The ever-changing coast is home to an unimaginable variety of life. It is no wonder that it is a magnet to people, drawn by its beauty, its resources and ability to inspire. Coastal zones are fragile, however, and need our protection to make sure they continue to provide all of these benefits into the future.

The next five chapters introduce different habitats found at the coast. They discuss why the different habitats are important and how people use them. The following chapters also highlight how people are changing coastal and marine areas and what is being done to manage this change, and in some cases protect marine habitats.

LEARN MORE

:: BBC Schools, Introducing the Coastline:
   www.bbc.co.uk/learningzone/clips/an-introduction-to-the-coastline/8429.html
:: BBC Schools, Rivers and Coasts:
   www.bbc.co.uk/schools/riversandcoasts/coasts/whatis_coast/index.shtml
:: Junior Farmer Field and Life School Modules on Fisheries and Aquaculture:
   www.fao-ilo.org/fao-ilo-youth/fao-ilo-jffls
:: National Geographic:
   science.nationalgeographic.com/science/earth/surface-of-the-earth/coastlines-article
:: World Ocean Review: worldoceanreview.com/en/wor-1/coasts

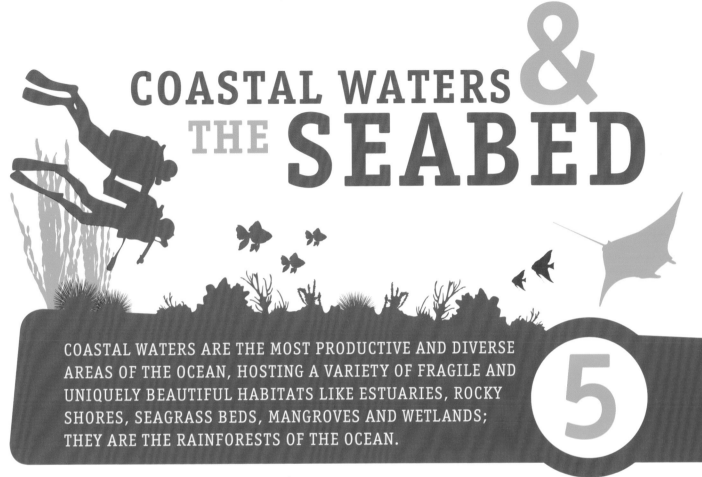

COASTAL WATERS & THE SEABED

COASTAL WATERS ARE THE MOST PRODUCTIVE AND DIVERSE AREAS OF THE OCEAN, HOSTING A VARIETY OF FRAGILE AND UNIQUELY BEAUTIFUL HABITATS LIKE ESTUARIES, ROCKY SHORES, SEAGRASS BEDS, MANGROVES AND WETLANDS; THEY ARE THE RAINFORESTS OF THE OCEAN.

Ana Queirós, Plymouth Marine Laboratory

Coastal waters and the seabeds beneath them are the shallow areas of the ocean and the ocean floor that represent the frontier between the land and the open ocean. These <u>habitats</u> occur close to land down to depths of 200 m, after which the <u>continental slope</u> occurs. These highly productive waters are occupied by large numbers of <u>species</u>, and provide a <u>habitat</u> for most marine fish during some stage of their life cycle.

CORAL REEF ILLUSTRATING THE DIVERSITY OF LIFE FOUND
ON THE SEABED.
© CBD

We can see some of these <u>species</u> close to the ocean floor (for instance when out snorkelling), but a very high number also live within the seabed, inside the <u>sediment</u>, such as cockles, flatfish, some lobsters and ragworms. These burrowing animals are an essential source of food for <u>species</u> like sea birds, otters, sea cows and even rays, playing a very important part in the marine <u>food web</u>.

BACKGROUND IMAGE
GANNETS DIVING AFTER FISH.
© Matt Doggett, Earth in Focus

BACKGROUND IMAGE (OPPOSITE PAGE)
SPOTTED EAGLE RAY.
© John Norton, WMC

CLOCKWISE, FROM TOP
A YOUNG PLAICE HIDING IN THE SEABED.
© Matt Doggett, Earth in Focus

SEA OTTER.
© Matt Doggett, Earth in Focus

CALIFORNIA SPINY LOBSTER.
© Shane Anderson, WMC

CRABS, BRITTLE STARS AND SEA URCHINS.
© Richard Shucksmith, Earth in Focus

NUTRIENT CYCLING AND FOOD WEBS

The shallow depth of coastal waters means that sunlight is able to reach organisms living in this part of the water column and the seabed. This presence of light, and the closeness to land, creates ideal conditions for the growth of organisms which, like the plants that live on land, depend on light to survive. These so-called photosynthetic organisms (from the Greek word 'photo' meaning 'light' and 'synthesis' meaning 'putting together') can be very small plankton and bacteria, as well as large seaweeds and seagrasses. They are the basis of the marine food web on which all other marine organisms rely. They feed the rest of the ocean!

PLANKTON.
© Claire Widdicombe,
Plymouth Marine Laboratory

SEAWEEDS.
© Toby Hudson, WMC

The growth of these important <u>photosynthetic</u> <u>organisms</u> depends on the availability of <u>nutrients</u>. Some of these nutrients come from rivers and the land, while others come from the deep ocean.

In healthy coastal waters, <u>nutrients</u> are quickly used up by <u>photosynthetic</u> <u>organisms</u> which grow, multiply and are eaten up by other <u>organisms</u>. Eventually all these <u>organisms</u> die, and their bodies gradually sink to the seabed, carrying with them all those used up <u>nutrients</u>.

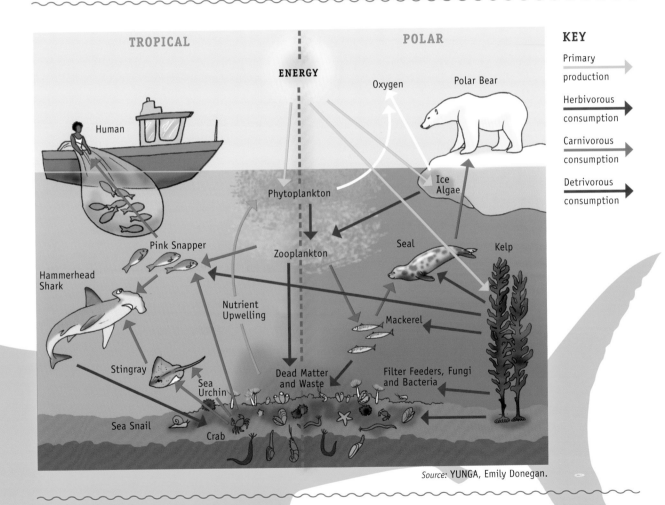

Source: YUNGA, Emily Donegan.

MARINE RECYCLERS

At the seabed (scientifically known as the benthos), small animals like worms, crabs, shrimps and clams live within the sediment, where they constantly burrow and move around in search of food. This burrowing moves the sediment around in a process known as bioturbation. Burrowing also mixes sediment particles and allows for bio-irrigation to take place; the movement of water into and out of the seabed.

THIS PACIFIC SAND CRAB (*EMERITA ANALOGA*) IS BURYING ITSELF. SOON, ONLY ITS ANTENNAE WILL BE STICKING OUT OF THE SAND INTO THE WATER, WHICH HELPS IT TO COLLECT FOOD.
© Jerry Kirkhart, WMC

BACKGROUND IMAGE
MOUNT EVEREST AS SEEN FROM A BHUTANESE AIRCRAFT.
© Shrimpo1967, WMC

? DID YOU KNOW?

These small burrowing animals are thought to move about 13 times the volume of Mount Everest across the whole sea floor every year!!!

THIS SNAPPING SHRIMP (*ALPHEUS BELLULUS*) AND YELLOW SHRIMP GOBY (*CRYPTOCENTRUS CINCTUS*) LIVE IN A SYMBIOTIC RELATIONSHIP: THE BLIND SHRIMP IS DIGGING A PROTECTIVE BURROW WHILE THE KEEN-EYED GOBY KEEPS WATCH.
© Nick Hobgood, WMC

These two processes – bioturbation and bio-irrigation – change the chemical environment of the seabed, making it ideal for very diverse communities of bacteria. The bacteria help break down the dead bodies of other marine life and any other organic matter that reaches the seabed. This recycling of organic matter makes nutrients available again to other marine life.

Burrowing organisms help return the nutrients back into the water from the sediment. The nutrients can then be taken up again by photosynthetic organisms, such as plankton and algae, taking us back to the beginning of the ocean food web...

This process works a bit like the bacteria that help break up the food in the human digestive system, making all the energy and nutrients in food available to our bodies.

Nutrient cycling (and recycling) is an essential part of a healthy ocean, without which most of the life in the ocean could not exist.

SPECIES THAT CREATE HABITATS

Some marine species provide a habitat for other marine species:

SEAWEED SHELTER:

Seaweeds sometimes grow in patches that are known as seaweed forests when they are big enough (e.g. kelp forests). Seaweeds provide shading for snails and limpets and many other marine creatures living on the seashore. It allows them to stay cool and hydrated during the hours of low tide. The seaweed is also a source of food for these and other marine creatures. As the tide comes in and the seaweed forests are submerged, they also provide food and shelter for many species of fish.

BIVALVE BEDS:

Some bivalves (e.g. mussels and oysters) live together in large beds and provide a home to many other species. Barnacles and other small organisms (e.g. worms and small crustaceans) live attached to the surface of mussels shells in mudflats and on rocky shores. The mussels also support the crabs and fish that eat them. You will also find animals hiding between mussel shells to keep out of sight of predators.

OTHER HABITAT ENGINEERS:

Other similar engineering species include other molluscs, corals, sponges, bryozoans, and any other species that can create a physical structure on the seabed when alive or dead: the hard surface provided by an empty shell, for example, can be the foundation of a new biological reef. Engineering species enrich the physical structure of the environment, helping to create the diverse patchwork of habitats that make up marine ecosystems. They are a vital part of the way in which the global ocean functions.

FISH SHELTERING IN KELP FOREST.
© John Turnbull, Flickr

MUSSEL BED EXPOSED AT LOW TIDE. CAN YOU SPOT THE LITTLE BARNACLES?
© kqedquest, Flickr

SCALLOP SHELL PROVIDING A HABITAT FOR BRYOZOANS.
© Richard Shucksmith, Earth in Focus

KELP FORESTS

Kelp is a species of marine algae largely found in temperate and polar waters, though some species inhabit tropical ocean regions. One of the most biodiverse and productive ecosystems on the planet, kelp forests provide a multitude of ecosystem services for marine life and people.

Kelp forests are often referred to as 'ecosystem engineers'. This is because as the kelp grows upwards, buoyed by small pockets of air in its leaves, it carries nutrients from the sea floor up through the water column. This, along with the safety and coverage provided by the canopy, makes kelp forests an ideal habitat for organisms such as fish. And where there are fish, there are also seals and sea lions!

Kelp forests also photosynthesize, providing marine life and people with oxygen, while absorbing carbon dioxide. As if this isn't a good enough reason to protect our kelp forests, they also make magnificent locations for scuba diving.

Because kelp forests provide shelter for small fish, this is a great place for large fish-eating species to find a meal! This concentration of fish around them means that kelp forests are, unfortunately, under threat from human activity such as overfishing. This is a good example of how interconnected food webs are: if humans eat too many carnivorous fish (or fish that eat other animals), the carnivorous fish eat fewer herbivorous sea creatures (plant-eaters), like sea

KELP FOREST.
© Stef Maruch, WMC

urchins. Without the large fish to prey on them, the herbivore community grows out of balance, overeating the kelp, and causing an imbalance in all the other species that depend on the kelp.

Action is being taken to protect kelp forests, though; for example, some have been designated as Marine Protected Areas (MPAs).

HUMAN ACTIVITIES:
IN, ON AND ABOVE THE SEABED

Coastal waters and the seabed are home to the largest diversity of marine species, because most of the food available in the marine food web is found in these parts of the ocean. Most of the fish, crustaceans and molluscs species that we like to eat, such as sardines, lobsters and cockles are found there. This wealth of marine life makes coastal areas desirable places for people to live.

COCKLES.
© Féron Benjamin, WMC

BACKGROUND IMAGE
SCHOOL OF SARDINES AT
MACTAN CEBU, PHILIPPINES.
© TANAKA Juuyoh, WMC

However, the results of human activities, like agricultural pollution of river waters, sewage discharge and coastal fishing activities that affect the seabed (e.g. trawling), can have negative impacts on coastal sea life.

We also use coastal waters and the seabed for much more than fisheries: for recreation, mariculture and as salt pans (for the extraction of aggregates like salt), as well as for laying cables and pipes. This intensive use of coastal waters means that human activity has destroyed more than 65 percent of sensitive habitats like seagrass beds and wetlands, and accelerated species invasions in this area of the ocean.

LAYING INSULATED OIL PIPELINES IN THE ARABIAN GULF.
© Shell, Flickr

FISH FARM IN HA LONG BAY, THAILAND.
© Gavin White, Flickr

BACKGROUND IMAGE
A FISHERMAN.
© www.sxc.hu

It is estimated that up to 80 percent of coastal seas are highly fished, to levels that cause serious concern.

MANAGING OUR **IMPACTS** ON **COASTAL** AREAS

MANAGING THE OCEAN

Our use and enjoyment of the coast depends on our ability to maintain clean and healthy coastal waters and seabed, and in recent years different strategies to reduce the impact of human activities on coastal habitats have been tried. These include governments giving coastal areas protected status, which necessitate the reduction or prevention of activities that would damage important habitats and threaten species. Measures to protect coastal habitats also include intervening on land, by, for example, controlling the discharge of pollution.

Individuals also have an important role in protecting their own beaches and coastlines. For example, voluntary efforts to clear litter from beaches are quite widespread, or groups of people may sign up to voluntary codes of conduct (ways of behaving) to reduce the negative impacts of seaside and marine recreation.

Perhaps you've heard the phrase *"Take only pictures and leave only footprints"*? What do you think it means?

THIS INNOCENT-LOOKING, FEATHER-LIKE ALGAE, CAULERPA TAXIFOLIA, HAS SPREAD FROM THE INDO-PACIFIC TO MANY OTHER SEAS, INCLUDING THE MEDITERRANEAN. IT REPLACES NATIVE PLANTS AND UPSETS LOCAL FOOD WEBS. LIKE THE LIONFISH, IT IS BELIEVED TO HAVE ESCAPED FROM AQUARIUMS.
© John Turnbull, Flickr

ORIGINALLY FROM THE INDO-PACIFIC, RED LIONFISH ARE GREEDY INVADERS OF U.S. AND CARIBBEAN WATERS. IN SOME AREAS, REWARDS ARE EVEN OFFERED TO DIVERS WHO SAFELY BAG AND REMOVE THE FISH FROM THE ECOSYSTEM THEY HAVE INVADED!
© FCW Fish and Wildlife Research Institute, Flickr

FACING PAGE
CURVED HULL WITH DEPTH MEASUREMENTS. THE BALLAST WATER IN SHIPS LIKE THIS MAY TRANSPORT MARINE ORGANISMS AROUND THE WORLD.
© Stephen Schauer, Lifesize/Thinkstock

HUMAN PRESSURES: INVASIVE SPECIES

The movement of ships has significant effects on the diversity of species around the globe. Ships on their long travels often take on stowaways or unexpected passengers.

This can happen in two main ways:
1. Species can be taken up in ballast water. Ships carry large volumes of seawater inside their hulls to maintain balance when low on cargo. This water gets emptied when a ship reaches port.
2. Marine species may colonize the hull of ships and live happily in this environment. Soft corals, sponges, and other organisms that feed on particles in the water column will thrive in this constant movement of water and can form dense colonies or groups.

When a ship reaches port these organisms have access to environments that they could not reach naturally. If the conditions are right, they might be able to survive there and ultimately change that ecosystem.

For instance, they might outcompete local species by using up all the space available, or consume all the food in the seawater, leading to a decline in the species that were naturally adapted to that location. In this case, the new species becomes 'invasive' and cause severe harm to the local biodiversity.

For some cosmopolitan species that have large worldwide distributions this is not problematic, but for others, particularly endemic species, it can even lead to extinction.

What is being done about it?

If you are familiar with boats, you may know that boat hulls should be treated with anti-fouling paint, to hinder the growth of unexpected passengers. This is one of the ways in which people try to minimize the likelihood of spreading invasive species.

Many efforts are also being made by international authorities to manage their spread. For example, the International Maritime Organization has put together a Ballast Water Convention. This treaty asks that the water carried in ship hulls must be treated before it is released at the destination port, preventing the arrival of 'unwanted' species. It is currently being successfully implemented around the world.

STUDYING THE SEABED:
SEDIMENT PROFILER IMAGERS

Understanding what goes on at and inside the seabed is challenging because things are happening underwater, and even when we can get to the seabed, things are happening inside the sediment as well! This means marine scientists have had to find new technologies that enable them to look inside these marine sediments, to observe processes while they are happening. One of these new technologies is called a sediment profiler imager (SPI). This complex invention works on the simple principle that to understand what goes on in the seabed we have to be able to look inside it!

MANUAL DEPLOYMENT OF PROFILE IMAGER DURING A BIOLOGICAL SURVEY IN THE VENICE LAGOON IN ITALY.
© Ana Queiros

PROFILE IMAGER, SITTING IN A FISHING BOAT, WAITING TO BE DEPLOYED.
© Ana Queiros

The <u>sediment profiler imager</u> is like a submarine <u>periscope</u>. Using a cunning system, a mirror is placed at an angle inside a <u>prism</u>. This can be pushed into the <u>sediment</u>, reflecting the image of a cross-section of the seabed to a camera, which takes a photograph for the scientists doing the research.

SPIs have allowed us to see what worms and crabs are doing below the surface of the seabed in their natural environment. Today, SPIs allow us to measure what is happening within the seabed in real time, not just to visualize activities in the seabed but also to quantify them. They are an essential tool in assessing the health of these <u>ecosystems</u>.

PROFILE IMAGES AS ACQUIRED ABOVE, SHOWING HEALTHY BROWN SEDIMENTS AT THE SURFACE OF THE SEABED, AND UNHEALTHY DARK SEDIMENTS FURTHER DOWN.
© Ana Queiros

A CRAB BIOTURBATING SEDIMENT BY MOVING AROUND FLORESCENT PLAY SAND ARTIFICIALLY ADDED TO THE SEABED.
© Solan *et al.* (2004) Marine Ecology Progress Series 271: 1-12

BACKGROUND IMAGE
DIVER SWIMMING NEAR THE SEABED.
© Comstock/Thinkstock

CONCLUSION

Coastal waters and the seabed are busy places – not only because of all the human activities going on there, but also because of the wealth of marine life found there. The seabed is alive with burrowing organisms, bioturbating and bio-irrigating, moving the sediment and water around the seabed, re-cycling nutrients. These nutrients are released back into the water, allowing photosynthetic organisms to grow, feeding the entire food web, including the fish and shellfish we like to eat. On the seabed you can find ecosystem engineers, species that maintain the chemistry of the seabed, and provide shelter and habitat to other species, supporting a diversity of marine life that we depend on, not only as a food source but as places we like to have fun in.

For these reasons, it is essential to keep coastal waters and the seabed clean and healthy. This maintains the health of the wider ocean, as well as the health of the goods that humans receive from it. One way this is being done is through the creation of Marine Protected Areas. International treaties are also being developed, for example, to reduce the spread of unwanted invasive species.

LEARN MORE

:: Ballast Water Convention:
www.imo.org/OurWork/Environment/BallastWaterManagement/Pages/Default.aspx

:: Bivalves:
www.molluscs.at/bivalvia

:: Invasive Species:

:: **www.blueocean.org/issues/changing-ocean/invasive-species**

:: **www.cbd.int/invasive/doc/marine-menace-iucn-en.pdf**

:: **www.iucn.org/about/work/programmes/marine/marine_our_work/marine_invasives**

:: **www.ocean.nationalgeographic.com/ocean/critical-issues-marine-invasive-species**

:: Kelp:
www.oceanservice.noaa.gov/facts/kelp.html

:: Ocean food chain video:
www.bbc.co.uk/nature/habitats/Neritic_zone#p003k0t0

:: Plankton video:
www.bbc.co.uk/nature/habitats/Neritic_zone#p00l28wc

THE HIDDEN WORLD OF ESTUARIES

HIDDEN WITHIN THE RICH MUDFLATS OF AN ESTUARY IS A WEALTH OF BIODIVERSITY THAT MAKES ESTUARIES AS PRODUCTIVE AS A TROPICAL RAINFOREST.

6

Jennifer Lockett, Plymouth Marine Laboratory

An <u>estuary</u> is formed where a river or stream runs into the sea forming a <u>transition zone</u> between <u>fresh water</u> and <u>saltwater</u>. An <u>estuary</u> is affected by both riverine influences, such as flows of <u>fresh water</u> and <u>sediment</u>, as well as marine influences, such as <u>salinity</u>, <u>tides</u> and waves. The mixing of <u>fresh water</u> and <u>seawater</u> creates a <u>brackish</u> environment where <u>salinity</u> is highest closer to the sea and lowest in the river.

AERIAL IMAGE OF THE EXE ESTUARY.
© Exe Estuary Management Partnership

Estuaries are generally characterized by their geographical features, often shaped by historical events such as glaciation. Different types of <u>estuaries</u> tend to be called different names, such as <u>fjords</u>, <u>rias</u>, and embayments (or <u>bays</u>), but the meeting and mixing of <u>fresh</u> and <u>saltwater</u> is what makes an <u>estuary</u> an <u>estuary</u>.

CLOCKWISE

AERIAL PHOTOGRAPH OF THE RÍO DE LA PLATA ESTUARY ON THE BORDER BETWEEN ARGENTINA AND URUGUAY.
© NASA

SAN FRANCISCO BAY IN CALIFORNIA, A SHALLOW, PRODUCTIVE ESTUARY THAT DRAINS WATER FROM APPROXIMATELY FORTY PERCENT OF CALIFORNIA.
© USGS

THE ESTUARY OF THE RIVER NITH, SCOTLAND, AT LOW TIDE; OPENING INTO SOLWAY FIRTH.
© Doc Searls, WMC

BACKGROUND IMAGE

GEIRANGERFJORDEN FJORD IN NORWAY.
© Simo Räsänen, WMC

DID YOU KNOW?

Each cubic meter of <u>estuary</u> mud contains the same amount of energy as 14 chocolate bars!

WHY ARE ESTUARIES SO RICH IN <u>BIODIVERSITY?</u>

Both <u>fresh water</u> and <u>seawater</u> carry <u>sediment</u> and <u>nutrients</u> in the <u>water column</u>. <u>Nutrients</u> in <u>fresh water</u> generally come from <u>run-off</u> from land, such as soil, silt and plant and animal <u>debris</u>. When the river opens up into the wider <u>estuary</u>, its flow decreases and much of this <u>sediment</u> is deposited, forming mudflats. Mudflats are characteristic of the upper reaches of an <u>estuary</u> at <u>low tide</u>. As salt from the <u>seawater</u> mixes with <u>fresh water</u>, it causes finer particles of <u>sediment</u> to clump together and sink – a process known as flocculation.

LIM KANAL IN CROATIA, AN EXAMPLE OF A RIA.
© Aconcagua, WMC

As <u>seawater</u> flows into an <u>estuary</u> it carries <u>nutrients</u>, sand and particles of marine <u>debris</u> (e.g. the dead bodies and faeces of marine <u>organisms</u>). Where this is deposited in the lower reaches of the <u>estuary</u>, sandflats are formed.

Deposited marine <u>debris</u> and <u>nutrients</u> start the intricate <u>food web</u> of an <u>estuary</u> by providing food for microorganisms, such as bacteria and <u>plankton</u>.

These in turn are fed upon by <u>invertebrates</u> including <u>crustaceans</u> (such as shrimps), <u>molluscs</u> (such as mussels) and <u>polychaetes</u> (worms).

The large quantities of <u>invertebrates</u> in many <u>estuaries</u> support vast numbers of wading birds, which feed on them at <u>low tide</u>. <u>Estuaries</u> are also important resting places for many <u>migratory</u> birds, providing them with food during their long journeys.

A WILSON PLOVER, A SMALL WADING BIRD.
© Richard Shucksmith, Earth in Focus

POLYCHAETE WORM.
© Matt Doggett, Earth in Focus

The size and shape of a bird's bill provides them with the perfect tool to target their food. In an <u>estuary</u> this can mean that different <u>species</u> of birds can forage together without competing. For example, plovers use short blunt bills to collect prey from the surface, while curlews have very long, curved bills to reach deep into the sand.

LONG-BILLED CURLEW.
© Alan D. Wilson, WMC

<u>Estuaries</u> provide more food and shelter than the open ocean, which is why many <u>species</u> of fish make use of this rich <u>habitat</u> during <u>juvenile</u> stages of their life cycle (i.e. when they are young).

Over two thirds of the fish and shellfish that we eat spend some part of their lives in an <u>estuary</u>.

HERRING EGGS ON WIREWEED.
© Kathleenreed, Flickr

BACKGROUND IMAGE
BIRDS ON A BEACH IN VENEZUELA.
© Emiliano Ricci, Flickr

THE INFLUENCE OF THE SEA

Some rivers meet the sea without forming an estuary because, for example, the geography of the river prevents seawater entering it (e.g. the land is too steep). In other cases, the river channel divides creating multiple, much smaller entry points to the sea. These are known as deltas.

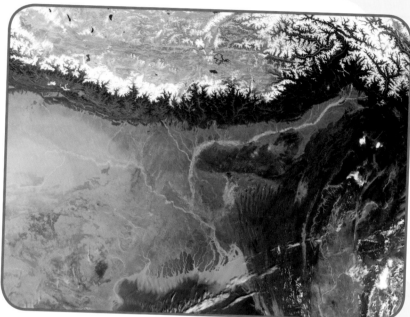

THE GANGES DELTA FROM SPACE.
© NASA, WMC

The world's largest delta is at the mouth of the Ganges River, which lies mostly in Bangladesh and covers an area of 105 000 km². The deposition of nutrients here has led to the area being one of the most fertile regions in the world.

The sea shapes the mouth of the estuary by sweeping sediment back out to sea with the tide and keeping the mouth of the estuary clear. If the sand is not swept away, it is likely to build up to form an alluvial plain or sand bar across the mouth of the estuary. As the sediment continues to build up, the alluvial plain may develop into a delta. Deltas often form when the sea at the estuary is shallow and/or calm for most of the year or where the gradient (steepness) of the river is very low or the river slows down significantly as it meets a body of water. This deposited sediment prevents the fresh water from flowing directly out to sea, creating many small channels through the alluvial plain that look like streams.

The sea's **tides** can also bring about unusal features in estuaries, such as the formation of **tidal bores**. These occur in just a few **estuaries** worldwide which have large **tidal ranges** and where incoming tides are funneled into a shallow, narrowing river. When the **tidal range** is at its largest (e.g. during the **spring tide**) the incoming water can appear as a sudden wave that travels rapidly upstream against the river **current**. This **bore** brings with it a rapid rise in water level which remains after the **bore** has passed.

DID YOU KNOW?

The Severn Estuary in the UK has one of the largest **tidal ranges** in the world. During the highest **tides**, the rising water is funnelled up the Severn Estuary into a **tidal bore** which has led to the development of river surfing.

SOME BRAVE PEOPLE SURFING THE FRONT WAVE OF THE SEVERN BORE.
© Tess, WMC

USING ESTUARIES

PEOPLE & OCEAN

Fish and waterfowl, reeds, seaweed, sand, clay and salt have been extracted from estuaries for thousands of years, providing rich resources for people who live along their shores. Before the use of road transport, even small estuaries were crucial transport hubs for the movement of people and goods. Fisheries and shipping services continue to this day, and a great variety of commercial uses have also developed as estuaries have become attractive places for recreation and tourism.

EXAMPLE USES AND BUSINESSES ASSOCIATED WITH AN ESTUARY:

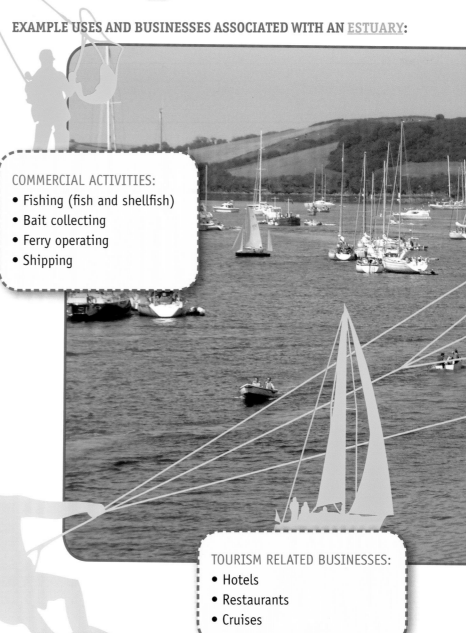

COMMERCIAL ACTIVITIES:
- Fishing (fish and shellfish)
- Bait collecting
- Ferry operating
- Shipping

TOURISM RELATED BUSINESSES:
- Hotels
- Restaurants
- Cruises

YOUTH AND UNITED NATIONS GLOBAL ALLIANCE

RECREATIONAL ACTIVITIES:
- Kitesurfing
- Sailing
- Windsurfing
- Canoeing
- Wildlife watching
- Dog walking

RECREATIONAL BUSINESSES:
- Mooring associations
- Boat yards
- Water sport schools and equipment hire
- Wildlife guides

SALCOMBE ESTUARY.
© Beverly Tremain

?
DID YOU KNOW?

Estuaries are incredibly important to people. Today, approximately 60 percent of the world's population lives along estuaries and coasts, and 22 of the world's largest cities are located on estuaries, including New York, London and Buenos Aires.

HOW IS
HUMAN ACTIVITY
AFFECTING BIODIVERSITY?

PEOPLE & OCEAN

Growing human populations have changed the boundaries around estuaries by developing harbours and marinas to support recreational and commercial activities, and reclaiming land for agriculture, construction and sea defences.

This has limited the habitat available to wildlife and brought it into closer contact with humans. For example, waders (long-legged birds) rely on the estuary for food during low tide. As our recreational use of estuaries increases, humans encroach more and more on these vital bird habitats, with activities ranging from kite surfing to dog walking.

Each time a bird is disturbed it will not only be prevented from feeding, but it is forced to use its precious energy reserves to fly away from the disturbance. If this happens frequently, it may eventually mean that the bird can't build up enough body fat to survive winter periods or long migrations.

THIS BIRD, A WILLET, HAS JUST CAUGHT ITS DINNER!
© Richard Shucksmith, Earth in Focus

OPPOSITE PAGE

BACKGROUND IMAGE
BARNACLE GEESE OVERWINTERING IN AN ESTUARY.
© Richard Shucksmith, Earth in Focus

THREE BARNACLE GEESE AT TÖÖLÖNLAHTI, FINLAND.
© Tomi, WMC

EUTROPHICATION AND THE IMPACT OF POLLUTION ON OUR ESTUARIES

In recent decades, agricultural practices, wastewater treatment plants, urban runoff, and the burning of fossil fuels have increased the level of nutrients (particularly nitrogen and phosphorus) entering estuaries, to many times the levels that occur naturally.

While some nutrient input is essential, these increased levels can be highly harmful. They often lead to the faster growth of algae. This process is known as eutrophication.

Excessive quantities of nutrients can result in dense algal blooms occurring in the estuary. These can become harmful if they grow densely enough to block sunlight to lower levels of the water column and prevent it from reaching photosynthesizing plants such as seagrasses on the seabed. Algal blooms may also use up the dissolved oxygen in the water, which is essential for the survival of other plants and animals. Some species of algae may even be toxic, and so may directly poison other marine life, including fish.

In severe cases, eutrophication can lead to a sharp loss of biodiversity and influence human activities such as fishing and recreation. Eutrophication can even be dangerous to human health, for example, if shellfish contaminated with algal toxins are eaten, or if people are directly exposed to waterborne or airborne toxins from the blooms.

WHAT IS BEING DONE TO HELP?

Estuaries are all different in terms of their physical and biological environment and are affected differently by the diverse uses made of them. For this reason, the local management of estuaries must be specific to individual sites, taking into account the needs of the local environment and the local community. Management measures must also look to the future to plan for changes that will impact upon the habitat, such as the need for sea defence structures to protect communities and infrastructure from sea level rise, and increased development to cater for a growing population.

National and International legislation, such as the Water Framework Directive (WFD) in Europe, is an important driver for improving our estuaries and setting targets for what we need to achieve. The WFD sets clear objectives and deadlines to improve water quality, but allows for local decisions on how this can best be achieved based on local knowledge, data and science.

At a global level we can all help to improve the health of estuaries by decreasing the amount of chemicals, such as washing detergents and pesticides, from entering our water systems.

Community-led initiatives can also be of great help to raise awareness of these issues. For example, yellow fish projects have been spreading around the world, using the motto: "Only rain down the drain!"

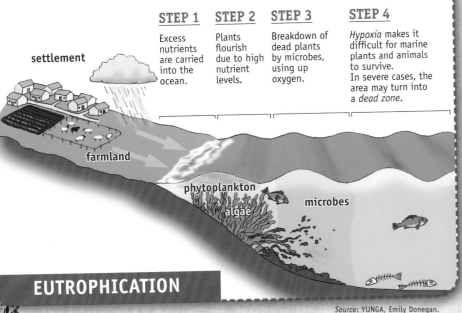

STEP 1
Excess nutrients are carried into the ocean.

STEP 2
Plants flourish due to high nutrient levels.

STEP 3
Breakdown of dead plants by microbes, using up oxygen.

STEP 4
Hypoxia makes it difficult for marine plants and animals to survive. In severe cases, the area may turn into a *dead zone*.

settlement

farmland

phytoplankton

algae

microbes

EUTROPHICATION

Source: YUNGA, Emily Donegan.

BACKGROUND IMAGE
ALGAL BLOOM IN A LAGOON.
© Dwight Burdette, WMC

MODELING ESTUARIES

The development of computer technologies has contributed greatly to our understanding of how estuaries 'work'. For example, scientists are now able to study the flows of water and predict how, for example, sediments and dissolved pollutants are likely to spread in an estuary over time. They are also being used to explore future flood risks. These estuary computer simulations or models can be helpful tools for estuary managers as they can identify where damage may occur before it actually happens, allowing for precautions to be taken.

ESTUARY MOUTH OF SANDY CREEK, AUSTRALIA,
AT LOW TIDE.
© Dwight Burdette, WMC

Getting these predictions right is an intricate business though. All kinds of factors need to be taken into account, including the properties of the sediments, as well as water speed, salinity, temperature and tidal cycle. And all of these predictions are checked against data collected from real-life studies, just to make sure the models are as accurate as possible.

CONCLUSION

Estuaries are special kinds of habitats, where salt water and fresh water mix. They are also special because they are food-rich environments that are important for a multitude of aquatic species, including many commercially important fish species, waders and waterfowl (such as ducks and geese). Estuaries are important for people too, who have used them for thousands of years for transport, fishing and recreation, among other things.

This human use continues to impact estuaries though, affecting the quality of the water found in them and the survival of species living there. While tailored local management is essential, we can all help to improve the health of estuaries by limiting our impact, decreasing the amount of chemicals and pollution released into our estuaries and being sensitive to valuable habitats, particularly when birds are feeding.

LEARN MORE

:: BBC Nature: **www.bbc.co.uk/nature/habitats/Estuary**

:: National Geographic:
www.education.nationalgeographic.com/education/encyclopedia/estuary/?ar_a=1

:: US Environmental Protection Agency: **www.water.epa.gov/type/oceb/nep/about.cfm**

:: US National Oceanic and Atmospheric Administration: **www.estuaries.noaa.gov**

:: Yellow Fish Initiatives:

:: **www.environment-agency.gov.uk/homeandleisure/pollution/water/120363.aspx**

:: **www.yellowfishroad.org/index.asp?p=2030**

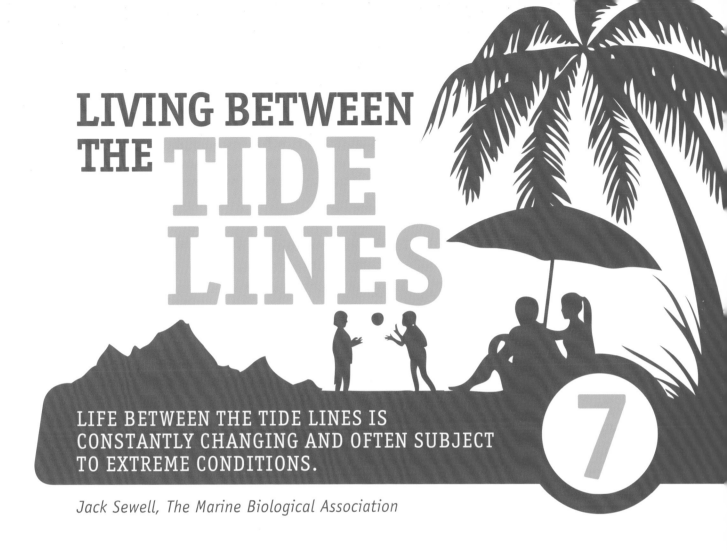

LIVING BETWEEN THE TIDE LINES

LIFE BETWEEN THE TIDE LINES IS CONSTANTLY CHANGING AND OFTEN SUBJECT TO EXTREME CONDITIONS.

Jack Sewell, The Marine Biological Association

7

The area between the <u>tide</u> lines is known as the <u>intertidal zone</u>. Marine creatures have had to develop a range of incredible adaptations and behaviours to survive here. When the <u>tide</u> retreats they find themselves exposed to the air and to great changes in their living conditions.

WAVE ON THE COAST OF MORRO BAY, USA.
© Mike Baird, Flickr

DIFFERENT
SEASHORES

Seashores provide very different types of <u>habitat</u>, because their characteristics can vary so greatly.

ROCKY SHORES provide a solid surface on which animals and seaweeds can fasten themselves. Open rock may be very exposed to the Sun, waves and predators, and therefore supports relatively few <u>species</u>, while extra hiding places such as under boulders, in pools, cracks, crevices and overhangs provide conditions suitable for a wider range of marine life.

A ROCKY SHORE.
© Jack Sewell

SEDIMENT SHORES can range from very fine mud through sand to coarse gravel. The material that forms the beach depends on the strength of the <u>tidal flow</u>, waves, wind and the amount of <u>organic matter</u> that reaches it. <u>Sediment shores</u> are more 'three dimensional' <u>habitats</u>, as animals are able to bury themselves beneath the shore's surface at varying depths. Conditions underground are more predictable and less extreme than above ground!

A SANDY BEACH.
© Nicholas Raymond, Flickr

SITTING IT OUT OR JUST VISITING?

Conditions don't only vary greatly between different types of shore, but also on different parts of a single beach.

Marine life on the lower shore is uncovered by the tide for a relative short time, so salinity and temperature levels are more stable and consistent. Competition for resources (such as space, food and light) between species on the lower shore is intense.

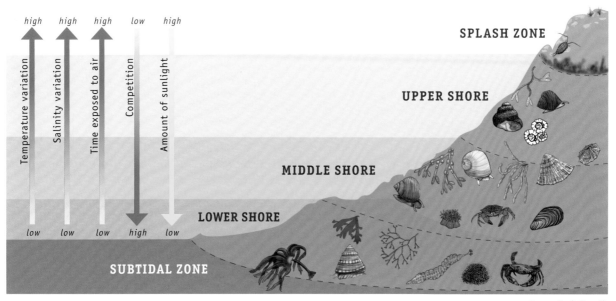

Source: Jack Sewell.

Conditions are much more variable and extreme further up the shore (see diagram above), but competition for resources becomes less. The ability of marine life to 'sit out' the times of exposure to air requires special behaviour or physical adaptations. Some species like the common cockle have adapted to absorb oxygen from the air when out of water but most marine species are unable to do this and must use other techniques to survive. Here are some examples:

BARNACLES

Some types of barnacle are excellent examples of <u>intertidal</u> adaptation. They form calcium plates to create a safe, cone-shaped refuge. When submerged, the barnacle uses its adapted legs to catch pieces of food passing its shell. When the <u>tide</u> retreats, the legs are pulled back in and plates cover the opening, trapping water inside. The barnacle cannot feed during this time, but can survive exposed for very long periods, enabling some to live at the very highest reaches of the <u>tide</u>.

BIVALVES

<u>Bivalves</u>, such as mussels and oysters, have a similar technique to barnacles, and use a strong muscle to keep their shell closed tightly and avoid drying out.

SEAWEEDS

Seaweeds can also adapt to the extreme conditions on <u>rocky shores</u>: A well-adapted seaweed <u>species</u> can spend 90 percent of its life out of the water and survive the loss of 80-90 percent of its moisture!

BLENNIES

More mobile <u>species</u>, such as fish, crabs and starfish, can retreat into rockpools, crevices and under boulders as the <u>tide</u> falls. Some blennies (a type of small fish) have adapted pectoral fins, which allow them to 'walk' out of the water and remain lodged in crevices! Their uncommon ability to <u>absorb</u> oxygen through their scale-less skin allows them to survive long periods out of the water.

BARNACLES.
© Jack Sewell

EDIBLE MUSSELS.
© Mark A Wilson, WMC

CHANNELLED WRACK, A COMMON BROWN SEAWEED.
© Jack Sewell

COMMON BLENNIE, ANGUILLA.
© Marc AuMarc, Flickr

IMPORTANCE TO **HUMANS**

PEOPLE & OCEAN

As we've mentioned, around 60 percent of the world's population live in coastal areas, drawn there by the valuable resources the ocean holds and the potential it offers for global transportation.

In some cold regions, the shore is one of the few areas which does not permanently freeze. This makes it a lifeline, supplying food through the winter months for <u>indigenous</u> people and animals.

Humans have exploited some of the <u>species</u> found on the coasts for centuries and now use the seashore to farm them on a large scale, although both wild harvesting and <u>aquaculture</u> can damage the natural <u>ecosystem</u> unless carefully managed. <u>Bivalves</u> such as cockles, clams and mussels are important food sources and support extremely valuable commercial harvests around the world. Seaweeds such as wracks and kelps are also used for food and in a range of cosmetic and industrial products.

Humans also use the seashore as a recreational area. Activities such as fishing, swimming, surfing and many others are undertaken on the shore worldwide. However, the development of tourism can have serious impacts on coastal life. Human activity can disturb birds, and light and noise from resorts may upset the habits of nesting turtles.

The importance of the ocean to people has led to the development of docks, housing and industry along the coastlines. Inevitably, this has caused problems for the environment, including loss of or damage to natural <u>habitats</u>, pollution and the introduction of <u>invasive species</u>.

A TYPICAL CROWDED BEACH IN TOSSA DE MAR, SPAIN.
© Katonams, WMC

YOUNG BOY WATCHING THE SEABIRDS ALONG THE SHORE, ON SIESTA BEACH (USA), AT SUNSET.
© Jsarasota, WMC

It is quite common for washed-up seaweed to be removed from beaches to make them more attractive to visitors. This can either be done by hand or by using mechanical beach cleaners. Mechanical cleaners are heavy pieces of machinery that can compact the sand and remove the top 10-15 cm of beach. This layer of beach is often rich in <u>nutrients</u> and provides <u>habitat</u> for

TRACTOR REMOVING VEGETATION FROM A A BEACH.
© Greg Henshall, WMC

shore <u>species</u>, meaning that beach cleaning is thought to remove up to 90 percent of the <u>species</u> found there.

Litter left by beach-goers and waste from tourist resorts can harm the shore and the animals living there as well.

BEDS OF HARDER WEALDEN SANDSTONE PROTRUDING
FROM THE SANDY BEACH BETWEEN HANOVER POINT AND
SHIPPARDS CHINE. ISLE OF WIGHT, U.K.
© Jim Champion, WMC

POLLUTION: OIL SPILLS AND THE TORREY CANYON

Jack Sewell and Annie Emery

On 18 March 1967, the Torrey Canyon oil tanker was wrecked on a rocky reef 17 miles from Land's End, Cornwall, England. The incident resulted in the release of 117 000 tonnes of crude oil into the sea. The Torrey Canyon is still one of the largest oil tanker spills in history, although it released only about 20 percent of the quantity of oil lost from the Deepwater Horizon spill into the Gulf of Mexico in 2010.

Oil spills on this scale can be damaging in several ways. Elements of the oil are toxic to marine and seashore creatures and make some commercially important species unmarketable by 'tainting' them. Crude oil is thick and sticky and can smother species, such as seabirds. Low-lying attached animals and seaweeds are also affected when their supply of light is block out and their feeding systems are blocked up.

The Torrey Canyon disaster triggered scientific research that has been crucial in guiding how oil spills are managed to this day. One key finding was that, whilst the oil had some immediate impacts, it was the detergents used to break up the oil which had the longest lasting and most catastrophic impacts on the ecology of the shore.

PREVENTION AND MANAGEMENT

Of course, the best way to manage oil spills and their devastating effects is to prevent them happening in the first place. This can be done by ensuring that oil tankers, machinery and other equipment are checked and repaired regularly to prevent leakages from occurring.

If an oil spill cannot be prevented, there are ways in which its effects can be reduced, including:

:: Having an action plan that all employees are familiar with and clean-up equipment ready for a speedy response to minimize harmful consequences.
:: Having an alternative tanker ready, so the oil can be transferred into this tank if the other begins to leak.
:: Containing the spilled oil, for example by pouring sand over the extent of the spill (which absorbs

the oil and then falls to the seabed in clumps that need to be removed), or by using containment booms (floating barriers which enclose the slick so the oil can be skimmed off the water's surface).

:: Re-using spilled oil – sometimes spilled oil can be collected and sent back to a refinery to produce usable oil again.

Volunteers are also incredibly important in the clean-up efforts following an oil spill.

Many groups are committed to saving sea birds and other animals harmed in oil slicks.

SMALLER LEAKS AND SPILLS

Oil spills from tankers and rigs often hit the media because they have such a severe, immediate impact. However, it is widely believed that oil spills from these accidents contribute only a small proportion of the oil entering the sea every year. A much larger source is thought to result from the sum total of small sources such as households and industry, which enters the sea from atmospheric emissions and run-off via drains. Day-to-day oil production causes smaller leaks and spills, which can severely affect the local environment if they are allowed to continue over a long period. One area that has been particularly badly affected by this kind of chronic oil pollution is the Niger Delta in West Africa.

You can make a difference by making sure you or your friends and family don't tip old oil into drains, and make an effort to catch and contain any leaks during machine care and maintenance.

DEEPWATER HORIZON DISASTER, GULF OF MEXICO.
© Justin Stumberg, U.S. Navy

MANAGING BEACHES

MANAGING THE OCEAN

We have seen that human activity can strongly affect – and harm – the ocean and ocean life. At the same time, natural processes may also impact on human activity: for example, the erosion caused by incoming waves, or, in more extreme cases, flooding, can completely reshape a coastline. As coastal areas are often highly developed, it is in our interest to make sure the ocean doesn't damage human settlements and infrastructure, or threaten lives and livelihoods.

Beach management strategies take two main approaches: 'hard' engineering options, which rely on permanent or semi-permanent structures, and 'soft' engineering techniques that are less permanent. Hard engineering includes the construction of seawalls that help waves bounce back to sea, groynes (wooden structures placed at right angles to the beach that catch sediments, preventing their erosion from the beach),

rip-rap (boulders placed at the bottom of cliffs to absorb wave energy and reduce erosion) and off-shore breakwaters (long structures, similar to rip-rap, that absorb wave energy and redirect waves). Soft engineering includes managed retreat, where any hard structures are removed and the coast is allowed to flood, and beach nourishment, where sand is brought to a beach to replace that carried away naturally by the waves.

A SEAWALL IN VENTNOR, ISLE OF WIGHT.
© Oikos-team, Wikipedia

GROYNES IN REDCAR, UK.
© Mattbuck, WMC

OPPOSITE PAGE
RIP-RAP IN CAPE TOWN, SOUTH AFRICA.
© Adam Brink, WMC

STUDYING THE SHORES

The seashore is more accessible than most marine habitats, as there is no need for expensive diving equipment or research boats when the tide is out. However, the tide does come back in, so scientists only have a short opportunity to count, measure, collect and observe shore species.

Consequently, the amount of information that can be collected is often limited. Many of the animals, plants and seaweeds 'shut down' when they are out of the water, making it difficult to directly study their behaviour. Scientists have therefore devised many ways of studying behaviour indirectly.

The low-tide period can be used to set up experiments and equipment on the shore.

For example, the effect of grazing behaviour on algal growth can be studied by setting up exclusion cages to prevent grazing snails from accessing certain areas of rock. Tagging and marking animals can also be used to follow their movements. Observing tracks, trails, grazing marks and egg cases are all additional ways a marine scientist can play detective and make a picture of what might take place when the sea returns.

TAGGING SEAWEED IN A SEASHORE STUDY.
© Jack Sewell

You can get involved in studying the seashore yourself! Next time you are on the shore, make a note and take photographs of the things you find and report them at **www.sealifesurvey.org**, which accepts records from all around the world. You will then be contributing information about what lives on the shore where and the information will be used by scientists and policy makers to help manage the shore in your area.

DID YOU KNOW?

CONCLUSION

The sea shore is an extreme environment. Different types of shore offer different challenges for marine life, but so does the location in which the organisms are found. The higher up the shore, the more extreme it becomes. The marine life found in this intertidal zone has to adapt to constantly changing conditions.

The sea shore is an important habitat for humans too, we particularly like to use it for leisure and recreation activities, but it also protects human settlements from erosion caused by waves and flooding during storms. Humans have therefore come up with many ways of managing our shores, some of which try to protect against the forces of nature (such as sea walls), but others allow nature to take its course (such as managed realignment of the coast).

LEARN MORE

:: Coastal management:

    :: www.geography.learnontheinternet.co.uk/topics/coastal_management.html
    :: www.georesources.co.uk/coastman.htm#L1

:: Marine Biological Association: **www.mba.ac.uk/education**

:: Marine Life Information Network: **www.marlin.ac.uk**

:: National Geographic : **www.ocean.nationalgeographic.com/ocean**

:: The Seashore.org: **www.theseashore.org.uk**

:: Simulate an oil spill clean-up:
www.education.nationalgeographic.co.uk/education/activity/simulate-oil-spill-cleanup/?ar_a=1

MANGROVES
AND SALTMARSHES

MANGROVES AND SALTMARSHES PROVIDE VITAL NATURAL BENEFITS TO PEOPLE, AND ACT AS BUFFERS BETWEEN LAND AND SEA.

8

Christi Turner, Blue Ventures Conservation
Tara Hooper, Plymouth Marine Laboratory/Marine Education Trust

Mangroves are trees that can tolerate high salinity.
Saltmarshes also contain many salt-loving plants. This means
that, unlike most plants, they can grow in the area bordering
the sea and the land: the intertidal zone.

MANGROVES IN LOS HAITISES NATIONAL PARK (DOMINICAN REPUBLIC).
© Anton Bielousov, WMC

MANGROVE ROOTS

Mangroves grow on mudflats in sheltered bays and inlets. This mud is often low in oxygen, so mangroves have characteristic roots that stick up above the sediment and allow the trees to take oxygen from the air.

When the tide comes in, the mangrove roots are immersed in salt water, although the tops of the trees are never submerged.

The different species of mangrove have different mechanisms for dealing with the salt. Some can prevent it from entering their roots, others store it in their leaves and other species secrete it from their leaves as salt crystals.

TOP
MANGROVES IN THE JUPITER INLET, USA.
© Kim Seng, Flickr

BOTTOM
SALT CRYSTALS ON AN AVICENNIA MARINA LEAF.
© Peripitus, WMC

BACKGROUND IMAGE
MANGROVES IN THE SALINAS ESTUARY IN SALINAS, PUERTO RICO.
© Boricuaeddie, WMC

THE TROPICS AND SUBTROPICS

The tropics are located in a band around the equator between the Tropic of Cancer in the northern hemisphere and the Tropic of Capricorn in the southern hemisphere. The subtropics are located directly north and south of the tropics, generally at a latitude of between 23.5° and 40° north and south. Moving even further away from the equator leads to cooler temperate zones, beyond which lie the cold polar regions at the Earth's highest latitudes.

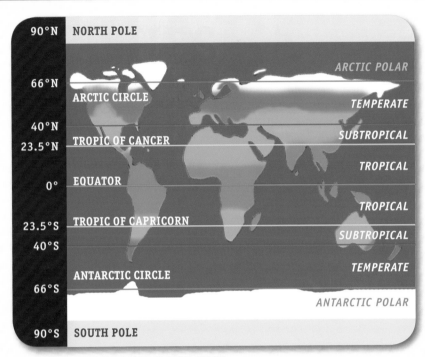

EARTH'S KEY GEOGRAPHICAL ZONES AND THEIR LATITUDES.
© YUNGA, Emily Donegan

:: Mangroves cover about 137 000 km² of the Earth's surface – an area bigger than the whole of Bangladesh!

:: Most of the world's mangroves are in the tropical and subtropical regions.

:: They are found in 123 countries around the world; however, 75 percent of the world's mangroves are found in just 15 countries.

:: Indonesia alone has almost a quarter of the world's mangroves, and Australia, Brazil and Mexico also have high densities of mangroves.

SALTMARSHES

Mangroves are not the only plants to fill this niche between the land and the ocean; in other areas saltmarshes fulfil this role. Saltmarshes look very different to mangroves, however, as they contain much smaller plants.

The main factor that divides regions with mangroves from regions with saltmarshes is temperature. Australia, Brazil, China and the USA are examples of countries that span subtropical and temperate zones, and so have both mangroves and saltmarshes, and a boundary between the two.

A SALT MARSH AT FORT FISHER STATE RECREATION AREA.
© Dincher, WMC

- As the <u>climate</u> gets colder, <u>mangrove</u> forests contain fewer and fewer <u>species</u> until it is too cold for any to survive, and <u>saltmarshes</u> take over.
- <u>Saltmarshes</u> seem to be less sensitive to temperature and can spread into <u>subtropical</u> (and even <u>tropical</u>) regions. But, in the <u>tropics</u>, <u>mangroves</u> are better at <u>colonizing</u> the environment than <u>saltmarshes</u>, partly because they are much taller and so shade the smaller <u>saltmarsh</u> plants, reducing their ability to grow.
- Sometimes <u>saltmarshes</u> and <u>mangroves</u> can be found in the same place, with the <u>saltmarsh</u> higher up the shore. This area is too salty for <u>mangroves</u> and is rarely <u>flooded</u> by the <u>tide</u>, but if <u>mangroves</u> are given the opportunity (for example through changes in sea level, land use or increased rainfall), then they will creep landward, replacing the <u>saltmarsh</u>.

SALT MARSH FORMING IN A TIDAL CREEK.
© Bill O'Brian, USDA

BACKGROUND IMAGE
SALT MARSH IN FLORIDA, USA.
© Bill Lea, U.S. Fish and Wildlife Service

WHY DO WE NEED
MANGROVES AND SALTMARSHES?

Mangroves and saltmarshes provide a range of what environmental scientists call ecosystem services: "the benefits provided by ecosystems that contribute to making human life both possible and worth living" (as defined by the UK Natural Ecosystem Assessment).

MANGROVE ISLAND, LAKE WORTH, USA.
© Kim Seng, Flickr

- Mangroves and saltmarsh plants filter and break down pollution from the land before it reaches and harms the marine environment. They also keep excess sediment from reaching sensitive habitats, such as seagrass beds and coral reefs, where it can smother the plants and corals or block access to light and restrict their growth.
- By reducing the energy of the waves that arrive at the shore, mangroves and saltmarshes act as a defence against storms and coastal erosion. It has been estimated that restoring mangrove areas in Vietnam saved US$ 6 million in coastal defence maintenance, even after allowing for the costs of replanting and protecting the trees.
- Mangroves and permanent ponds within saltmarshes provide hatching and growing areas for a variety of species of fish and crustaceans (like crabs or shrimp), offering them protection for the early stages of their lives before they head to the open ocean. Larger fish use saltmarsh creeks and the marsh flats at high tides as feeding grounds.
- Wading birds and wildfowl use saltmarshes for roosting (resting or sleeping), and feed in the nearby mudflats. Mangroves also support migratory and resident birds populations, including birds of prey such as ospreys, which hunt from, and nest in, the trees.

PROTECTING MANGROVES

Only about 7 percent of the Earth's mangroves are currently being protected – that's only about 9 505 km² or about the size of the small Indonesian island of Buru.

As mangroves, seagrass beds and coral reefs are all closely linked, scientists recommend that conservation efforts should protect entire corridors, combining protection for mangroves, seagrass beds and coral reef at the same time.

Mangrove conservation projects have been most successful when the local community is involved, in an approach known as community-based conservation. This means that, while preserving the mangrove environment is a top priority, it is also essential that the local community is involved as much as possible and that the conservation activities improve their quality of life.

An example of young people taking a leading role in the protection of mangroves is the Binapani Youth Club of Naranpur village in India, where members, called 'Mangrove Guardians', take turns protecting the mangrove forest. Their responsibilities include tending to the community mangrove nursery and educating the community in the important role of mangroves in the ecosystem.

THE MANGROVE GUARDIANS OF THE BINAPANI YOUTH CLUB.
© Centre for Coastal Sustainability (CCS), India

MANGROVE, SIAN KA'AN ECOLOGICALRESERVE, MEXICO.
© Claire Murphy, Flickr

115

STUDYING MANGROVES AND SALTMARSHES

Scientists use a technique called remote sensing to monitor the extent of mangrove forests and saltmarshes. This involves analysing satellite images of an area to determine where different types of habitat and land use can be found. It is even possible to distinguish between different types of mangrove forest, because the characteristics of the different species (for example, Rhizophora and Avicennia) affect the way the trees reflect light.

Remote sensing has to be 'ground-truthed', which involves collecting data manually in the mangrove forest or saltmarsh to check what each part of the satellite image represents. This can be challenging for mangrove areas, because they are not the easiest environment to study: mangrove forests tend to be very dense. Their trees have tangled roots in thick mud and they are often in inaccessible locations.

Remote sensing images can be combined with other information to create a Geographical Information System (GIS). This produces a map showing the mangrove or saltmarsh and surrounding area, which is overlain with additional information to show, for example, where towns and villages, roads, shrimp ponds, and fishing areas are located. Remote sensing and GIS techniques can be applied to other habitats too, including the surface of the ocean (see p.172).

MANGROVE FORESTS SATELLITE IMAGE. BOMBETOKA BAY, MADAGASCAR.
© NASA

CONCLUSION

Mangroves and saltmarsh plants can survive the salty conditions found in the intertidal zone. While mangroves are found only in the tropics and subtropics, saltmarshes are found across the world. Both mangroves and saltmarshes provide people with many ecosystem services. They play a role in the removal of pollution, they protect the coast against storms and erosion and they provide important habitat for young fish, and species such as shrimps and crabs.

Mangroves and saltmarshes also help with the storage of carbon dioxide, preventing it from entering the atmosphere. This role is under threat, however, as both mangroves and saltmarshes are disappearing because of human activities. Very little of the world's mangroves and saltmarshes are protected, but community activities are making a difference. For example, people are replanting mangroves and educating their communities about their importance.

LEARN MORE

:: The Mangrove Action Project: **www.mangroveactionproject.org**
:: Mangrove Guardians: **www.apowa.org.in/case-studies-2/mangrove-guardians**
:: Mangrove Watch:
   www.mangrovewatch.org.au/index.php?option=com_content&view=frontpage&Itemid=300165
:: The Seashore.org:
   **www.theseashore.org.uk/theseashore/Saltmarsh%20section/succession%20general/
   The%20saltmarsh%20environment.html**
:: WWF: **www.wwf.panda.org/about_our_earth/blue_planet/coasts/mangroves**

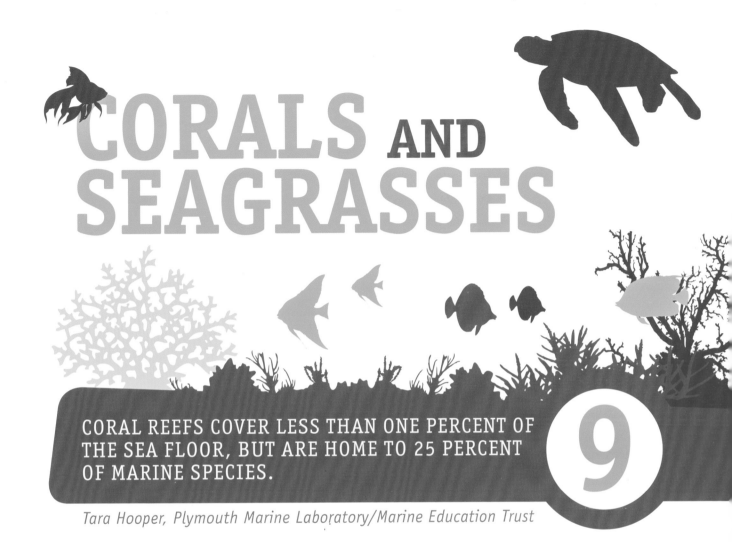

CORALS AND SEAGRASSES

CORAL REEFS COVER LESS THAN ONE PERCENT OF THE SEA FLOOR, BUT ARE HOME TO 25 PERCENT OF MARINE SPECIES.

9

Tara Hooper, Plymouth Marine Laboratory/Marine Education Trust

Corals are colonies of tiny individual animals called polyps. Coral polyps have soft bodies and stinging tentacles, just like the anemones to which they are related. In tropical corals, the polyp is protected by a hard, external skeleton made of calcium carbonate. These skeletons grow into a range of shapes, some form branching structures or delicate fans, others look a bit like a human brain!

CORAL REEF IN FIJI SHOWING A DIVERSITY OF CORAL SPECIES.
© Julie Bedford, NOAA PA

119

TROPICAL CORAL REEFS

Warm water corals are found in the shallow, sunlit zones of tropical and subtropical oceans. These corals use their tentacles to catch their prey of microscopic animals in the water called zooplankton. However, most of a coral's energy is produced by microscopic plants called zooxanthellae that live within the coral. These zooxanthellae actually give corals their beautiful colours.

GREAT BARRIER REEF FROM SPACE.
© NASA

Where many corals grow together, they are said to form a <u>reef</u>. There are two main types of <u>reef</u>, <u>barrier reefs</u> and <u>fringing reefs</u>.

- <u>Barrier reefs</u> are found some distance from the shore, and are usually older structures in comparatively deep water. Probably the most famous coral <u>reef</u> is Australia's Great Barrier Reef, which extends for 2 000 km along the coast of Queensland.
- <u>Fringing reefs</u> are found in shallower water, often close to the coast. They can completely surround an island, creating a large enclosed lagoon. An <u>atoll</u> is a special form of <u>fringing reef</u> that no longer surrounds any land; the island has sunk beneath the waves while the coral has continued to grow upwards.

Not all corals build reefs, however. There are more than 2 500 species of coral, of which only about 650 are reef-builders.

ATOLL, MALDIVES ISLANDS.
© WMC

OENO ISLAND AND ITS FRINGING REEF.
© NASA

BACKGROUND IMAGE
BRANCHING CORAL.
© CBD

?
DID YOU KNOW?

Corals are often very slow growing: it can take massive corals 50 years to grow to the size of a football! Branching corals grow more quickly, but their brittle branches are easily broken. The combination of being easily damaged and slow to repair makes coral reefs very vulnerable to human activities.

COLD WATER
CORALS

Corals are not just found in warm waters, some species can live in the cold and darkness at depths of up to 4 000 m. In deep water there is not enough light for algae to survive, so cold water corals do not host the zooxanthellae that are characteristic of tropical reef-builders.

TUBASTREA MICRANTHA, A TYPE OF SOFT CORAL.
© Matt Doggett, Earth in Focus

Our knowledge about cold water corals is very limited. It was only in the 1970s that scientists really began to learn more about them, as fisheries and oil and gas exploration expanded into deeper waters.

We do know that cold water corals are a very important within deep water **ecosystems**. The deep sea is mostly a flat expanse of muddy **sediments**. The **reefs** formed by cold water corals are complex, three-dimensional structures, which provide a **habitat** for other marine life. Over 1 300 other **species** have been found living on a single cold water **reef**.

A SOFT CORAL KNOWN AS A TREE CORAL, RED SEA, EGYPT.
© Matt Doggett, Earth in Focus

A PINK SEA FAN.
© Matt Doggett, Earth in Focus

A COLD WATER CORAL KNOWN AS 'THE BUBBLEGUM CORAL'.
© NOAA, WMC

123

SEAGRASSES

Seagrasses get their name from their long ribbon-like leaves. They are not grasses, although they are flowering plants. In fact, they are more closely related to land plants than to seaweeds.

Seagrasses are much less diverse than corals, but much more widespread. There are only about 60 species of seagrass, but they are found in the shallow coastal waters of every continent except Antarctica. In tropical areas, they are often found within the lagoons created by coral reefs.

Seagrasses range widely in their size too, from small species measuring just 2-3 cm, to giant varieties with leaves over 4 m long. They are an important food source for green turtles, sirenia (see box) and some species of water fowl including black swans.

GREEN SEA TURTLE GRAZING SEAGRASS AT AKUMAL BAY.
© P.Lindgren, WMC

Sirenia are a group of large aquatic <u>herbivores</u>, which grow up to 3 m long. They can be found in rivers, <u>estuaries</u> and shallow coastal seas. The group comprises a single <u>species</u> of dugong, found in the Indian Ocean and eastern Pacific, and three <u>species</u> of manatee, which inhabit the southeastern United States, the Caribbean, South America, and West Africa.

Dugongs in particular are very dependent on seagrass for their food, and an adult can eat up to 40 kg of seagrass a day! Healthy seagrass beds are critical to dugong survival; even before a lack of seagrass leads to starvation, dugongs do not breed when their food is in short supply, which can have serious consequences for endangered populations.

A DUGONG IN MARSA ALAM, EGYPT.
© Julien Willem, WMC

BACKGROUND IMAGE
A SEAGRASS MEADOW.
© Matt Doggett, Earth in Focus

USING SEAGRASSES AND REEFS

SNORKELING IN BALI.
© WMC

People make fairly limited direct use of coral, fragments of which may be sold as ornamental curios or be shaped into jewellery. However, coral reefs support almost a quarter of all marine fish species, as well as a huge variety of other species including crabs, lobsters, shellfish, and octopus.

Seagrass has wider uses, particularly as carpeting and matting, but also as mattress stuffing, packing material and for roofing and insulation. Seagrass beds also play a role as a habitat for young fish and crustaceans, which take shelter beneath the leaves.

Coral reefs and seagrass beds are therefore vital in the production of food: up to one billion people eat seafood derived from coral reefs, and thirty million people are thought to be entirely dependent on them.

Other ecosystem services we gain from coral reefs and seagrass beds include:

- Protection of the shoreline from storm damage;
- Reduced coastal erosion because the roots of seagrass plants stabilize the sediment;
- The coral sand that creates and replenishes beaches; and
- The opportunity to watch fish, rays, sharks, turtles and the other wealth of life on tropical reefs, which supports the huge global tourism industry.

As some kinds of coral have a chemical composition very similar to human bone, they have been used as an alternative source of bone grafts, helping damaged human bones to heal!

DID YOU KNOW?

A BAG MADE OF SEAGRASS.
© WMC

THREATS TO CORAL REEFS AND SEAGRASS BEDS

Both coral reefs and seagrass beds are fragile environments and are vulnerable to human activities.

It is estimated that almost one third of seagrass habitats have disappeared since the nineteenth century. The rate of loss has increased in recent decades; since 1980, an area of seagrass about the size of Belgium has been lost!

The extensive fishing that takes place on coral reefs can do major damage. Trampling, anchors, heavy fishing gears and, most destructively, fishing with dynamite, all do significant damage to the fragile corals. Activities on land also threaten reefs: soil run-off can prevent light from reaching the zooxanthellae and can smother the coral, and pollution can be toxic.

There are similar threats to seagrass beds, which are easily ripped up by the careless anchoring of ships. The excess nutrients in the water that result from eutrophication can also destroy seagrass, which may be choked by blooms of algae that result.

PROPELLER SCAR IN SEAGRASS.
© Sarah Manuel

A CORAL ROCK THAT HAS UNDERGONE DAMAGE CAUSED BY HUMAN ACTIVITIES.
© NOAA

THE IMPACTS OF CLIMATE CHANGE

Burning coal, oil and gas for heat, electricity and transport has increased the amount of carbon dioxide and other greenhouse gases in the atmosphere. This is leading to a rise in global temperatures, known as climate change. This creates two particular problems for coral reefs: increased ocean temperature and an increase in ocean acidity.

Rising temperature

Rising temperatures worldwide also affect the temperature of the ocean. Tropical coral reefs in shallow water are particularly vulnerable to this. If water temperatures are just 1-2 °C higher than their usual summer maximum for more than about three weeks, the corals lose their zooxanthellae and ultimately die.

When a coral loses its zooxanthellae it also loses its colour and turns white. This is called coral bleaching. The frequency and severity of coral bleaching has been increasing. The most serious mass bleaching event of recent years occurred in 1998 and caused widespread coral death across the world's reefs. More localized mass bleaching events occurred in 2002 on the Great Barrier Reef and in 2005 in the Caribbean.

Seagrass is also affected by increased seawater temperature. The impacts can occur quickly; just a few hours of increased temperature can reduce photosynthesis and "burn" seagrass leaves a brown-black colour, indicating that the plant's cells have been damaged. High water temperature is thought to have been a factor in the loss of large areas of seagrass in Florida and Australia.

Increasing acidity

The other problem for corals is that about a quarter of the extra carbon dioxide entering the atmosphere ultimately ends up being absorbed by the ocean. This makes the seawater more

↗

BLEACHED CORAL.
© Elapied, WMC

acidic, in a process called ocean acidification. As the water's acidity increases, the concentration of a substance called carbonate in seawater decreases. Carbonate is essential for many marine organisms, including corals, as they use it to form the calcium carbonate that makes up their skeletons and shells. Not all corals are affected in the same way, but it is thought that reduced carbonate concentrations will slow down the growth of many corals. Corals that are still able to develop are likely to have less dense skeletons that can be damaged more easily.

It has been predicted that the combined effects of coral bleaching and ocean acidification could result in the extinction of one third of reef-building corals, severely reducing the habitat reefs provide to so many other species.

Seagrasses, on the other hand, may actually benefit from a higher concentration of carbon dioxide in the ocean. It will encourage more photosynthesis, allowing seagrasses to grow more vigorously. However, many of the organisms that live on seagrasses, such as coralline algae, are expected to be negatively affected by ocean acidification, leading to a reduction in the diversity of organisms found living in seagrass beds.

BACKGROUND IMAGE
RED FERN CORAL.
© CBD

PROTECTING REEFS:
THE EXAMPLE OF THE GREAT BARRIER REEF

The Great Barrier Reef is the world's largest coral reef and lies in the Coral Sea off the coast of Queensland, Australia. It is spread over around 344 400 km², which is almost the size of Germany! The best thing is, the Great Barrier Reef is alive! In fact, it is the only living thing on Earth that is visible from space. Among its 3 000 individual coral systems live 1 500 different species of tropical fish, 200 bird species and 20 types of reptile, including the endangered green sea turtle.

Not only is the Reef a rich and productive habitat, but it also has great economic importance, with tourism generating over US$ 4 billion every year. Therefore, protecting the Great Barrier Reef is essential – without it we would lose the ecosystem services that it provides on a massive scale.

To do so, nearly all of the Great Barrier Reef has been designated as a marine park, which is a type of Marine Protected Area (MPA). More and more MPAs are being created around the world to protect coral reefs, and other important habitats, by restricting the activities that can take place on and around them.

To maintain a balance between protecting the reef and still being able to use it, the Great Barrier Reef marine park is divided into zones. Within the highly protected zones, which now cover one third of the park, the only activities that are allowed are: scientific research; recreational diving and boating; traditional use by indigenous people; tourism; and the passage of ships. Most of these activities are controlled by a permit system.

These highly protected zones are balanced by general use zones, which cover a further third of the park, and in which a wide range of activities, including commercial fishing are allowed, mostly without permits.

↗

In between these extremes of protection are zones in which some of the most damaging activities (such as bottom trawling and large net fishing) are prohibited, while activities which have a lower impact (including line fishing and crabbing) are permitted.

While the creation of Marine Protected Areas is growing, protection of the marine environment still falls behind protection of terrestrial environments. In 2013, only half of one percent of the marine environment was fully protected in no-take marine reserves. More than 15 times as much land receives this sort of protection.

THE SALOMON ATOLL IN THE CHAGOS ARCHIPELAGO (THE WORLD'S LARGEST MARINE RESERVE WITH AN AREA OF 647 497 KM^2).
© Anne Sheppard, WMC

CORAL **SURVEYS**

As we have seen, coral <u>reefs</u> are extremely important, but also extremely fragile <u>ecosystems</u>, meaning that scientists need to take special care when studying them. The survey data they collect can help us decide how best to manage coral <u>reefs</u> and whether they need extra protection. The questions scientists ask about <u>reefs</u> include:

REEF RANGERS CARRYING OUT A CORAL REEF SURVEY.
© Matt Doggett, Earth in Focus

Mapping: How much sea floor does the <u>reef</u> cover? And how dense is the reef?
To answer these questions, scientists may use sophisticated <u>satellite</u> remote sensing imagery – or simply go snorkelling! For example, the 'Manta Tow Technique', which has been used widely around the Great Barrier Reef, involves the snorkeler taking notes on waterproof paper while being towed along by a motorboat!

<u>Ecosystem</u>: What species are found on the reef? How do they interact with one another?
Again, a snorkeler or diver is best for this form of data collection, though without the boat this time. This allows the swimmer to reach greater depths and explore more delicate areas of the reef.

Human activity: What impacts do humans have on the <u>reef</u>? How can the local community be involved in preserving the <u>reef</u>?
Once the human impact on the <u>reef</u> has been studied (e.g. by survey, interviews, etc.), educational programmes are a good way to encourage the local community to help protect the coral <u>reef</u>, for example by explaining government rules and regulations, and demonstrating how the community can use the <u>reef</u> sustainably.

CONCLUSION

Corals come in many shapes and sizes and are found in shallow waters of the warm tropics and temperate areas, as well as in the cold deep ocean. They are fragile organisms that are very susceptible to the impacts of human activities. Seagrasses are less diverse, but are more widespread. They are also affected by human activities, but they may actually be one of the winners of climate change.

Marine Protected Areas are being created around the world to protect habitats such as coral reefs and seagrass beds. Some of these protected areas include zones where no human activities can take place together with zones where limited activities can occur. Unfortunately Marine Protected Areas are relatively few and much needs to be done to provide the ocean with effective protection.

LEARN MORE

:: The Coral Reef Alliance: **www.coral.org/resources**

:: Discover coral reefs:
www.kidsdiscover.com/spotlight/coral-reefs

:: The Great Barrier Reef: **www.greatbarrierreef.org**

:: ReefCheck: **www.reefcheck.org**

:: Restoring coral:
www.oceantoday.noaa.gov/coralrestoration

:: The World Wildlife Fund: **www.wwf.panda.org/
about_our_earth/blue_planet/coasts/coral_reefs/**

:: Seagrass Watch: **www.seagrasswatch.org/home.html**

:: UNESCO on mapping coral reefs:
www.unesco.org/csi/pub/source/rs12b.htm

:: Ocean acidification animation: the other CO2 problem:
www.youtube.com/watch?v=55D8TGRsl4k

Section

C

OCEAN EXTREMES:
WHAT MANY OF US WILL NEVER SEE

Chapter 10
BEYOND SIGHT: THE EXTREMES OF THE OCEAN

Chapter 11
THE FROZEN OCEAN

Chapter 12
THE SUNLIT ZONE

Chapter 13
THE DEEP SEA TREASURE CHEST

BEYOND SIGHT: THE EXTREMES OF THE OCEAN

MOST OF US WILL NEVER VISIT THE MAJORITY OF THE OCEAN – SO LET'S EXPLORE IT'S EXTREMES IN THIS GUIDE!

10

Caroline Hattam, Plymouth Marine Laboratory

Most of us only experience the ocean when we go to the coast, or perhaps if we take a ferry journey. We may just get a glimpse of its vast expanses in a documentary on the television. Even modern day seafarers, such as explores or people who work at sea, will only ever see a small fraction of the ocean.

VIEW OF EARTH'S HORIZON AS THE SUN SETS OVER THE PACIFIC OCEAN.
© NASA

EXPLORING
THE OCEAN

Humans have visited less than 10 percent of the ocean. This means that we know very little about the other more than 90 percent, which is still unexplored! While the technology exists to send people to the Moon, much of the equipment needed to explore beyond the coastal zone is still in development. This is because reaching the bottom of the ocean is not unlike travelling to the Moon – it requires a lot of specialist technology and an awful lot of money! It was only in 2012 that the technology was available for the first solo descent to the bottom of Challenger Deep (the deepest point of the ocean) and you can find out more about this in Chapter 13.

A GRAPHIC RENDITION OF THE SEAORBITER.
© Jacques Rougerie, SeaOrbiter

This may begin to change with the construction of floating oceanographic laboratories, such as the SeaOrbiter. Like an ocean-going space station, the SeaOrbiter will have laboratories and living space under the water's surface. It will allow scientists to spend extended periods under the ocean's surface to learn more about life and conditions there.

MODERN OCEAN EXPLORERS

Humans have been ocean explorers for thousands of years, as we learnt in Chapter 3, but there are some notable recent ocean explorers who have changed the way with think about the oceans:

Jacques Cousteau
(1910-1997) oceanographer, film-maker and underwater explorer: he made expeditions to the Mediterranean, the Red Sea and the Indian Ocean and brought the ocean to people's living rooms through more than 120 television documentaries and 50 books. He also founded the Cousteau society which aims to help protect the ocean from the dangers of human activity.

Jacques Piccard
(1922-2008) oceanographer and engineer: he designed and developed underwater vehicles for studying the ocean and, together with Lt. Don Walsh, was part of the first manned exploration to the bottom of Challenger Deep. They surprised the scientific community when they observed fish and shrimps at the bottom of the ocean, as scientist thought no life could survive the pressure at this depth. This discovery led to the prohibition of the dumping of nuclear waste in ocean trenches.

Sylvia Earle
(1935-present) oceanographer, aquanaut, author and former chief scientist of NOAA (National Oceanic and Atmospheric Administration, USA): she has led over 70 marine expeditions including more than 6 500 hours underwater. In 1970 this included the Tektite 2 project where she and four other women spent two weeks underwater at a depth of 15 metres, living in a specially designed 'underwater habitat' from where they could study ocean life and the effects of living underwater on the human body. In 1979 she also walked for two and a half hours at 381 metres underwater in a pressurised metal suit, similar to a space suit, with only a communication line connecting her to a submersible.

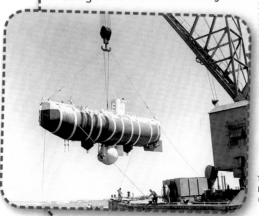

TRIESTE, ONE OF THE MOST IMPORTANT DEEP-DIVING RESEARCH BATHYSCAPHES.
© US Navy

SUPPORTING **LIFE** ON EARTH

Much of the ocean we don't see is cold (and some of it frozen), unimaginably deep, and a long way from land. These ocean extremes are exceptionally important to life on Earth and therefore also human existence. They provide a multitude of <u>ecosystem services</u>.

THE FROZEN OCEAN.
© Nat Wilson, Flickr

The frozen oceans of the Arctic and Antarctic affect the <u>climate</u> of the entire planet, influencing the currents of the <u>Thermohaline Circulation</u> that transport heat around the Earth. The ocean's ice sheets also provide unique <u>habitats</u> for marine life such as microorganisms, fish, mammals and birds.

The thin sunlit zone of the upper ocean is home to millions upon millions of microscopic life forms that contribute to the oxygen in our atmosphere. The wind-driven <u>surface currents</u> in this ocean zone help to transfer <u>nutrients</u> and life around the planet. This zone is also important for shipping and fisheries.

The vast deep sea is central to the recycling of <u>nutrients</u> that supports life in the rest of the ocean. It is being explored as a source of new resources, such as minerals, oil and gas, and deep sea fisheries. Biotechnology companies (companies that use living <u>organisms</u> to create useful products) are also interested in the deep sea as the <u>organisms</u> living there could provide solutions to medical and engineering problems.

HOW **PEOPLE AFFECT** THE EXTREMES

> **Despite the fact that most of these ocean extremes have never been visited by humans, the impact of human activities is felt throughout the ocean.**

For example:

- Plastics are particularly problematic and are found on the seabed and in the <u>water column</u> from the Arctic to the Antarctic.

- Many pollutants, such as DDT (a chemical compound used as an insecticide on farms) and heavy metals, are found throughout the marine environment, even in Arctic sea ice.

- Fishing is altering the structure of marine <u>food chains</u> by removing <u>species</u> and it is changing fragile marine <u>habitats</u>, such as deep sea coral <u>reefs</u>, <u>seamounts</u> and <u>sponge</u> fields.

- The impact of deep sea resource extraction is unknown, but scientists are concerned that mining activities could harm <u>habitats</u> about which we know very little.

FIRE BOAT RESPONSE CREWS BATTLE THE BLAZING REMNANTS OF THE OFFSHORE
OIL RIG, DEEPWATER HORIZON, ON APRIL 21, 2010.
© US COAST GUARD

PLASTIC SANDWICH BAG FLOATING IN THE WATER COLUMN. FISH THAT
FEED ON VARIOUS SALPS, JELLYFISH, ETC. MISTAKE SUCH GARBAGE
FOR FOOD AND MAY TRY TO EAT IT WITH FATAL CONSEQUENCES.
© Ben Mierement, NOAA

Who takes responsibility for these vast, distant areas of ocean is constantly in question. There are international **Conventions** that define the rights and responsibilities of countries using the world's ocean, such as the **UN Convention on the Law of the Sea**, but people's use of the sea is changing. International legislation and regulations therefore need to adapt to prevent the overexploitation of resources and the protection of **species**, **habitats** and other undiscovered treasures.

MARINE DEBRIS. A FORMER CORAL REEF STREWN WITH RUBBISH.
© David Burdick, NOAA

CONCLUSION

It is not just the coastal ocean that we need to take care of, but the global ocean as a whole. Managing the ocean, however, is challenging because no one country has responsibility for it: countries, communities and individuals need to work together for a common goal.

The following chapters introduce the frozen ocean, the sunlit zone and the deep sea. They explain why these ocean extremes are important to humans, how we are using them and how scientists are trying to learn more about these challenging environments.

LEARN MORE

:: BBC Nature: **www.bbc.co.uk/nature/habitats/Deep_sea**
:: BBC One Frozen Planet series: **www.bbc.co.uk/programmes/b00mfl7n**
:: MarineBio The Open Ocean: **http://marinebio.org/oceans/open-ocean**
:: WWF Oceans: Threat and Management:
   wwf.panda.org/about_our_earth/teacher_resources/webfieldtrips/oceans_threat

THE FROZEN OCEAN

THE POLAR OCEANS ARE THE FROZEN FUEL CELLS OF OUR PLANET, WITH UNIQUE ECOSYSTEMS AND CLIMATES.

Helen Findlay, Plymouth Marine Laboratory

11

The Arctic Ocean in the northern hemisphere and the Southern Ocean surrounding Antarctica in the southern hemisphere together make up nearly 10 percent of the global ocean. Over 72 percent of these polar oceans are covered by sea ice.

ICEBERG IN NORTHEAST
GREENLAND NATIONAL PARK.
© Rita Willaert, Flickr

WHAT IS <u>SEA ICE?</u>

Sea ice forms, grows and melts in the ocean; it is simply frozen <u>seawater</u>. <u>Perennial</u> sea ice, or <u>multi-year ice</u>, is believed to have existed in the Arctic for at least the last 700 000 years and could have been present for up to 4 million years!

Sea ice grows and shrinks each year. In the Arctic Ocean, sea ice reaches its maximum extent in March, extending as far south as Bohai Bay, China (38 ºN) in the Pacific, and as far as Iceland in the Atlantic Ocean. During spring and summer the sea ice melts, reaching its minimum extent in September when it covers only the main Arctic Ocean <u>basin</u> (see graphic below).

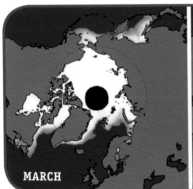

MARCH **SEPTEMBER**

MAXIMUM AND MINIMUM SEA ICE COVER FOR THE ARCTIC (TOP) AND ANTARCTIC (BOTTOM).
© NSIDC, University of Colorado, Boulder, Colorado

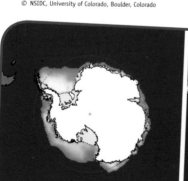
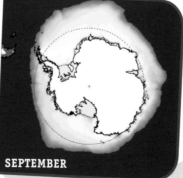

MARCH **SEPTEMBER**

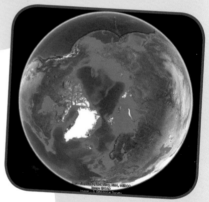
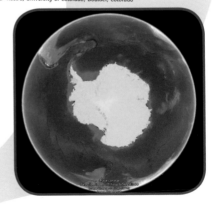

THE SATELLITE VIEW ON TOP SHOWS THE NORTHERN HEMISPHERE FOCUSING ON THE ARCTIC OCEAN, WHILE THE VIEW ON THE BOTTOM SHOWS THE SOUTHERN HEMISPHERE FOCUSING ON THE SOUTHERN OCEAN SURROUNDING ANTARCTICA.
© NSIDC, University of Colorado, Boulder, Colorado

In contrast to the Arctic Ocean, the Southern Ocean (surrounding the Antarctic continent) has no, or very little, perennial sea ice. The sea ice in the Southern Ocean forms and melts each year. At its maximum, it extends well beyond the Antarctic Circle (see the figure on p.146).

Sea ice forms and melts seasonally as part of a natural cycle driven by the change in the position of the Sun through the year. In polar regions, sunlight hours decrease through autumn until there is 24-hour darkness in winter, air temperatures can drop to well below -40 °C, which causes the surface seawater to freeze.

As the sunlight returns in spring, temperatures begin to rise and the sea ice begins to melt. New ice, or first-year ice is ice that forms each year and tends to be between a few centimetres to two meters thick, while the ice that remains over the summer in the Arctic (multi-year ice) can build up to over four meters in thickness.

SEA ICE IN THE ARCTIC.
© Helen Findlay, Plymouth Marine Laboratory

WHY IS <u>SEA ICE</u> IMPORTANT?

THE ALBEDO EFFECT

Sea ice is important for regulating the climate. Sea ice has a bright surface, which reflects much of the sunlight that hits it back into space. The 'reflectiveness' of a surface is known as its albedo. As a result of sea ice having high albedo, areas covered with sea ice don't absorb much of the Sun's heat (solar energy), so temperatures in the polar regions remain relatively cool.

If temperatures keep getting warmer due to climate change, and greater and greater amounts of sea ice melt, there will be fewer and fewer bright surfaces available to reflect the solar energy back into space. More of this heat will be absorbed by the dark oceans, which have low albedo, causing temperatures to rise further and starting a cycle of ever-increasing warming and melting.

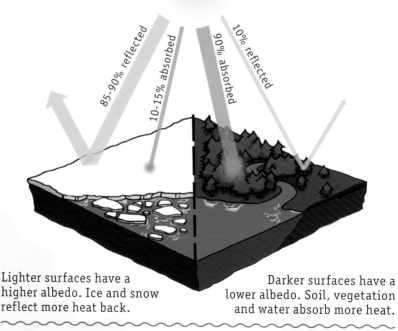

HEAT

85-90% reflected

10-15% absorbed

90% absorbed

10% reflected

Lighter surfaces have a higher albedo. Ice and snow reflect more heat back.

Darker surfaces have a lower albedo. Soil, vegetation and water absorb more heat.

Source: YUNGA, Emily Donegan

THE THERMOHALINE CIRCULATION

Sea ice contributes to the ocean's global circulation system or the <u>Thermohaline Circulation</u>. When sea ice forms, most of the salt in the water is pushed out of the ice into the ocean below. Water below the sea ice has a higher concentration of salt and is <u>denser</u> than the surrounding <u>seawater</u> and so it sinks.

In this way, sea ice affects the movement of ocean waters (see graphic below): as cold, <u>dense</u>, polar water sinks and moves along the bottom of the ocean towards the <u>equator</u>, warm water from mid-depths rises to the surface and travels from the <u>equator</u> towards the poles.

Without the <u>Thermohaline Circulation</u>, countries close to the <u>equator</u> would be far too hot and countries close to the poles would be far too cold for humans to survive. As well as cycling heat, the water <u>currents</u> caused by the <u>Thermohaline Circulation</u> are also vital for moving <u>nutrients</u> and carbon around our planet.

THE *GULF STREAM*

In the North Atlantic, the <u>current</u> of warm water flowing along the surface is called the Gulf Stream. By carrying warm water towards Europe, it has a big impact on the local temperature. For example, the annual average sea surface temperature off the coast of Northern Europe is about 12 °C. At the same <u>latitude</u> off the American east coast, annual average sea surface temperatures are only around 3 - 4 °C.

Changes in the amount of sea ice that forms and melts each year can disrupt these normal circulation patterns, which in turn influence the <u>climate</u> we experience, particularly in Northern Europe and America.

DIAGRAM OF THE THERMOHALINE CIRCULATION.
© Canuckguy, WMC

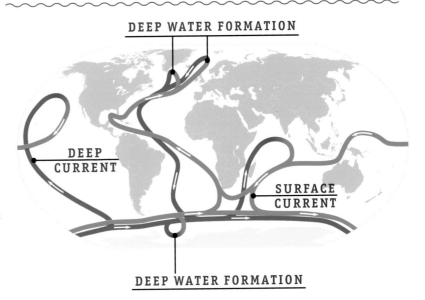

DEEP WATER FORMATION

DEEP CURRENT

SURFACE CURRENT

DEEP WATER FORMATION

WHAT LIVES ON AND IN SEA ICE?

Sea ice is an ecosystem that provides a habitat for many large animals, such as polar bears, seals and walruses, but also supports a huge diversity of microscopic plants and animals. These animals have adapted to survive in the extreme Arctic environment. For example, polar bears, arctic foxes and snowy owls have white fur or feathers to provide camouflage, helping them to catch their prey. Polar bears in fact have black skin under their white fur, because it absorbs more heat! The walrus has a thick layer of blubber to preserve its body warmth – fully-grown males can weigh up to two tonnes; that's as heavy as a car!

KING PENGUINS (*APTENODYTES PATAGONICUS PATAGONICUS*), WEST FALKLAND.
© Ben Tubby, Flickr

LARGE WALRUS ON THE ICE.
© Captain Budd Christman, NOAA

SEALS IN THE LEMAIRE CHANNEL, ANTARCTICA.
© Liam Quinn, Flickr

ICE ALGAE.
© Kils and Marschall, WMC

Sea ice isn't only home to large animals, though: microscopic plants called ice algae live in the channels which develop in sea ice as it forms, when the salt water is pushed out. These algae are grazed upon by microscopic animals such as copepods and other zooplankton.

Ice algae may fall from the ice into the water column, where even more animals will feed on them. This food source gets passed up the food chain to fish, seals, polar bears, whales and, of course, humans. Some of the algae dies and falls to the sea floor, where it is eaten by animals that live on the seabed such as clams, crabs and lobsters. Turn back to p.61 for an illustration of a polar food web.

A FEMALE POLAR BEAR AND HER CUBS.
© Travel Manitoba, Flickr

PEOPLE AND
THE FROZEN OCEAN

The frozen ocean is an important part of human culture. For centuries, <u>indigenous</u> people have lived in the frozen lands above the Arctic Circle, mostly concentrated near the coast. These people rely on sea ice for transportation and hunting.

AN INDIGENOUS FAMILY IN THE ARCTIC.
© Ansgar Walk, WMC

Indigenous people have a unique culture that is associated with the ice and snow. They have a complex language that describes ice structures and processes, which has allowed them to survive in this extreme environment. This detailed knowledge of sea ice includes an understanding of how the ice has changed and is changing. Many traditional stories and songs, passed from parent to child through multiple generations, record the changes that have taken place long before scientific records began, which is very useful for modern-day climate scientists!

While no humans have ever lived permanently in Antarctica, the Southern Ocean has played an important role in providing resources for humans for hundreds of years. Fur seals have been hunted for their pelts, elephant seals and penguins for their oils and whales for their fat – which was used to make products like soap!

Unfortunately, some of these species were hunted almost to the point of extinction. Global campaigns, such as Greenpeace's Save the Whale campaign, have aimed to ban hunting these endangered Antarctic species for sport, and in 1994, the International Whaling Commission designated the waters of the Southern Ocean around Antarctica as a whale sanctuary in which commercial whaling is not allowed.

HUMAN ACTIVITIES IN THE SOUTHERN OCEAN

1790: Sealers (seal hunters) first began hunting fur seals for their pelts.

1825: Some fur seal populations were close to extinction. Elephant seal and penguin hunting begins.

1900s: Whaling in the Southern Ocean begins. Finfish, crab and squid fishing in the region increase.

1970s-1980s: Southern Ocean krill fishing increases dramatically. Krill are small crustaceans that feed on plankton. They are an important source of food for whales, penguins, seals, squid and fish.

1990s: Krill catches drop, partly because of the introduction of quotas to encourage the sustainable exploitation of krill, and have remained stable ever since. Today most krill is used in aquaculture as feed, but also as fish bait, or for food for livestock and pets.

ANTARCTIC KRILL: *EUPHAUSIA SUPERBA*.
© Uwe Kils, WMC

THE ANTARCTIC TREATY

The **Antarctic Treaty** is one of the most successful international agreements on environmental protection. It was signed in Washington on 1st December 1959 by twelve countries whose scientists had been active in and around Antarctica during the International Geophysical Year (IGY) of 1957-1958.

The treaty entered into force in 1961, and has since been signed by many other nations (it currently has 50 members).

The treaty aims to protect the land and ice shelves south of the 60 °S latitude. In 1964, **Agreed Measures for the Conservation of Antarctic Fauna and Flora** were adopted. These measures provided a system of general rules and regulations that gave extra protection to Specially Protected Areas.

MOUNT HERSCHEL (3 335 M HIGH), ANTARCTICA. CAN YOU SPOT THE SEABEE HOOK PENGUIN COLONY IN THE FOREGROUND?
© Andrew Mandemaker, WCM

THE NEW ELEVATED AMUNDSEN-SCOTT SOUTH POLE STATION. THE CEREMONIAL SOUTH POLE IS IN THE FOREGROUND. THE FLAGS OF THE ANTARCTIC TREATY'S ORIGINAL 12 SIGNATORY NATIONS ARE ON DISPLAY BEHIND IT.

© U.S. Antarctic Program, National Science Foundation

Other important provisions of the treaty include:

- Antarctica shall be used for peaceful purposes only;
- Freedom of scientific investigation in Antarctic and cooperation shall continue;
- Scientific observations and results from Antarctica shall be exchanged and made freely available.

The Antarctic Treaty did not initially include the surrounding seas (the Southern Ocean), but scientists became concerned that unregulated increases in krill catches in the Southern Ocean could be negatively affecting Antarctic marine ecosystems. This led to the **Convention on the Conservation of Antarctic Marine Living Resources** (CCAMLR), which was adopted in 1980.

This convention works alongside the **Antarctic Treaty**, as well as the **International Convention for the Regulation of Whaling** and the **Convention for the Conservation of Antarctic Seals**, to protect and manage the marine resources south of the Antarctic Convergence. This is the region encircling Antarctica (at about 55°S) where cold waters flowing away from the continent meet, and sink beneath, the relatively warmer waters further north.

STUDYING THE
FROZEN SEAS

The frozen seas are one of the most challenging marine environments for scientists to study. The presence of sea ice and the harsh conditions – freezing cold, wind and storms, and dry conditions – prevents scientists from easily accessing the Arctic and Southern Oceans. There are two main ways that scientists get around this:

NY ÅLESUND, SVALBARD, A CENTRE FOR
INTERNATIONAL SCIENTIFIC RESEARCH ON THE ARCTIC.
© Karen Tait, Plymouth Marine Laboratory

A TEMPORARY SCIENCE BASECAMP IN THE ARCTIC.
© Martin Hartley, Catlin Arctic Survey

- In summer, and when there is less ice present, research ships can be used. Sometimes these ships are supported by 'ice-breaker' ships. These ice-breakers are designed to smash through <u>first-year sea ice</u>, but even these reinforced ships can get stuck! The ships can moor up alongside ice floes allowing the scientists to work on and around the ice.

- In other areas, scientists operate out of science bases.

These are camps, either temporary or semi-permanent, that are set up in specific, scientifically interesting locations. In the Antarctic, many countries have their own science base. Some scientists and support staff (which include electricians, chefs, builders, etc.) just work from the bases in the summer, but often a few people are left to live and work there all year round. There are also several bases in the Arctic,

but these are predominantly owned and run by countries that border the Arctic Ocean (like Canada, Russia, Norway and Greenland/ Denmark). When people 'over winter' in either the Arctic or Antarctic, they have to endure the 24 hour darkness, temperatures dropping well below -40 °C, and wind speeds over 70 mph. They may also face dangers in the form of wild animals, such as polar bears in the Arctic...

TWO ICEBREAKERS.
© United States Coast Guard

SEA ICE MELT IN THE ARCTIC AND GLOBAL WARMING

Since the introduction of satellites in 1979, the area covered by sea ice each season (also known as sea ice extent) has been monitored closely. Before this time, the best records for sea ice extent came from whaling ships and polar expeditions dating back to the 1800s. By monitoring the sea ice extent through time, scientists have been able to observe how the sea ice has been changing in the Arctic Ocean as the planet warms.

Between 1979 and 2014 there has been a large decrease in the area covered by sea ice in September (the time of year when there is least ice cover); the September sea ice extent minimum has decreased by 13.3 percent per decade.

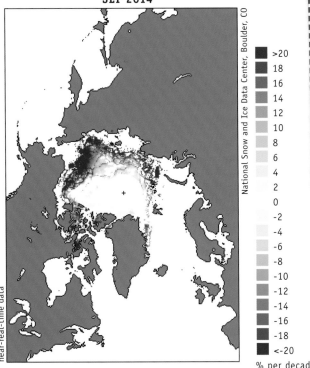

SEA ICE CONCENTRATION TRENDS SEP 2014

National Snow and Ice Data Center, Boulder, CO

near-real-time data

% per decade

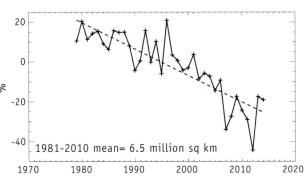

NORTHERN EMISPHERE EXTENT ANOMALIES SEP 2014

1981-2010 mean= 6.5 million sq km

slope= -13.3(+/-2.8)% per decade

FIGURES SHOWING CHANGE IN ICE EXTENT (ABOVE) AND THICKNESS (OR CONCENTRATION) REDUCTIONS (LEFT).
© NSIDC

BACKGROUND IMAGE
TASIILAQ GREENLAND IS AN ICY WONDERLAND INHABITED BY TEAMS OF SLED DOGS, COLOSSAL GLACIERS, AND HOME TO THE WORLD'S SECOND LARGEST ICE SHEET.
© Christine Zenino, WMC

There has also been a decrease in the sea ice thickness and a loss of the thicker <u>perennial</u> sea ice. If this trend continues, the overall scientific opinion is that the Arctic Ocean will be generally ice-free in the summer within next 30 to 50 years.

WHAT WILL THIS MEAN?

Less sea ice will mean fewer <u>habitats</u> are available, which could severely damage the Arctic <u>food web</u>. Human communities that rely on the sea ice for hunting will encounter difficulties if the sea ice melts too quickly in spring and shortens the hunting season. Local communities will need to adapt to new ways of living rapidly, and could risk losing many traditions.

GREENLAND'S GLACIERS DUMPING ICE INTO THE ATLANTIC OCEAN.
© Tim Norris, Flickr

Ice-free summers also mean industry and commercial organizations will look to exploit new resources. The oil and gas industries are already looking for new reserves, fishing fleets are looking for new fish <u>stocks</u> and shipping industries are looking for shorter shipping routes.

The benefits from the wealth of new resources will come at an environmental cost as the Arctic will remain a challenging and hostile environment, increasing the risk of damage and disasters during the winter and periods severe <u>weather</u>.

CONCLUSION

The frozen oceans of the Arctic and Antarctic play an extremely important role in the regulation of the global climate. By reflecting sunlight back into space, the ice keeps the polar regions cool. As the seawater around the ice is very salty, this dense, cold water sinks, contributing to the circulation of deep ocean currents.

Scientists are concerned that global warming could lead to the melting of both Arctic and Antarctic sea ice and even ice-free summers in the Arctic. This will affect polar food webs and will force human communities dependent on the ice for hunting to adapt. In attempt to protect the Antarctic, the Antarctic Treaty was developed as well as a number of conventions. Maybe it is time for an Arctic Treaty?

LEARN MORE

:: Association of Polar Early Career Scientists (APECS):
www.apecs.is/outreach/polar-outreach-publications/polar-resource-book

:: Digital Explorer Frozen Ocean Resources:
www.oceans.digitalexplorer.com

:: International Polar Year:
www.ipy.org/links-a-resources

:: National Snow & Ice Data Centre (NSIDC):
www.nsidc.org/cryosphere/education-resources

:: Royal Geographical Society (with IBG):
www.rgs.org/OurWork/Schools/Teaching+resources/Teaching+resources.htm

:: WAGGGS' Arctic Wildlife Fact File:
www.wagggsworld.org/en/grab/23595/1/wildlife-factfile-v5.pdf

THE SUNLIT ZONE

MOST LIFE IN THE OCEAN CAN BE FOUND NEAR THE SURFACE WHERE SUNLIGHT HELPS PHYTOPLANKTON TO GROW.

12

Frances Hopkins, Plymouth Marine Laboratory

Although the ocean is vast and has an average depth of 3 720 metres, most life in the ocean is found in a very thin layer at the surface. This is because light from the Sun can only penetrate seawater to a maximum of about 200 metres. Amazingly, a single cupful of seawater may contain billions of viruses, millions of bacterial cells, hundreds of thousands of phytoplankton and tens of thousands of zooplankton.

SUNLIT ZONE IN GUANTANAMO BAY.
© Shane Tuck, U.S. Navy

TINY LIFE FORMS
WITH A HUGE IMPORTANCE

These sunlit layers of the ocean are home to phytoplankton – tiny single-celled plants, barely visible to the naked eye. But although small in size, they have an enormous impact on our planet! Despite the fact that oceanic phytoplankton make up only about 1 percent of all biomass on Earth, they perform nearly 50 percent of all photosynthesis!

If phytoplankton are the grass of the ocean, then zooplankton are its cows. Zooplankton are the tiny animals that feed on the phytoplankton. They are crustaceans, like crabs and lobsters, but are much smaller in size. Zooplankton are food for many other sea creatures, including everything from the fish and shellfish that we like to eat, up to the mighty whales.

ZOOPLANKTON.
© Matt Wilson, Jay Clark, NOAA

PHYTOPLANKTON.
© Gordon T. Taylor, NOAA

YOUTH AND UNITED NATIONS GLOBAL ALLIANCE

PEOPLE AND PLANKTON

PEOPLE & OCEAN

As well as forming the base of the ocean food web, the phytoplankton in the sunlit ocean perform other important roles, including helping the ocean to capture and store huge quantities of carbon dioxide (CO_2).

When **phytoplankton photosynthesize**, they draw CO_2 down from the atmosphere. This is very good for us humans, as it reduces the amount of this **greenhouse gas** in the atmosphere. However, as humans continue to release huge quantities of CO_2, it's not such good news for the ocean. This is because human emissions of CO_2 are changing the chemistry of the ocean's surface: as we found out on pp.128-129.

When CO_2 is **absorbed** by the ocean, it makes the **seawater** slightly more **acidic**. This process of **ocean acidification** is happening at a rapid rate and marine scientists are concerned about how it will affect **organisms** living in the ocean.

SCUBA DIVER.
© Pachango, Flickr

THE OCEAN'S
MOTIONS

The ocean is constantly in motion. Huge ocean currents transfer massive amounts of heat energy around the Earth, bringing warm water from the equator to the higher latitudes, and carrying cooler water back towards the equator. There is more heat energy in the top 2.5 meters of the ocean than in the whole of the atmosphere.

Currents are also vital for the food chain and biodiversity, as they distribute nutrients, fish eggs and larvae across the ocean. The Humboldt Current that flows north from Antarctica along the east coast of South America supports huge fish stocks, which in turn attract sea birds, mammals, and humans: the area supports some of the world's largest fisheries.

MAP OF THE OCEAN CURRENTS.
© WMC

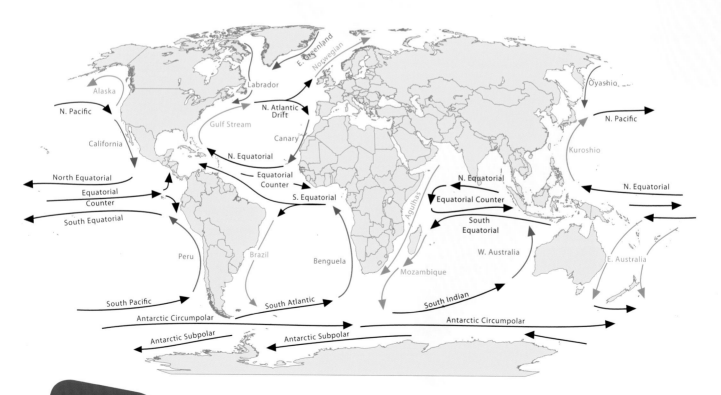

CURTIS EBBESMEYER, AN OCEANOGRAPHER WHO HAS CAREFULLY STUDIED THE MOMENTS OF THE FRIENDLY FLOATEES.
© Rick Rickman, NASA

DID YOU KNOW?

In 1992, a container ship on its way from Hong Kong to the USA lost its <u>cargo</u> of 29 000 rubber ducks (and beavers, frogs and turtles) during rough <u>weather</u>. These little plastic toys have proven to be invaluable in helping us to understand more about our ocean <u>currents</u>. In the 22 years since they were lost, the plastic creatures have washed up on shores all over the world, including Hawaii, Alaska, South America, Australia, and Scotland. Some have even made it across the North Pole, travelling encased in ice through the Arctic Ocean!

The map below shows the journeys made by different groups of the Friendly Floatees. By tracking and mapping the progress of the rubber toys, oceanographers have been able to predict or 'model' where more of them might appear. Thanks to this accidental science experiment, we now have a greater understanding of our ocean <u>currents</u>!

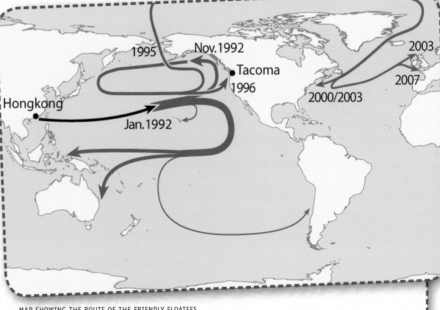

MAP SHOWING THE ROUTE OF THE FRIENDLY FLOATEES.
© NordNordWest, WMC

THE GREAT PACIFIC GARBAGE PATCH

Surface ocean currents are driven by the winds, and the directions in which they blow is fairly constant and predictable. Where you get two winds blowing in opposite directions over the ocean, large current loops form. These are called gyres.

By nature, gyres draw in debris from the surrounding ocean and slowly accumulate material – and humans have put a lot of rubbish in to the ocean! Once such debris gets trapped within a gyre, it's unlikely to get out of it again for a long time. The gyre in the North Pacific Ocean is infamous for the large amount of plastic waste that has accumulated in it, and is now known as the Great Pacific Garbage Patch. Not only is plastic waste in the ocean unsightly, it also has harmful effects on wildlife. Animals that feed at or near the surface of the ocean can accidentally ingest plastic and other waste when they are feeding. Sea turtles often mistake plastic bags for jellyfish and marine birds such as albatrosses may get caught in the plastic rings from fizzy drinks packaging. Others accidentally feed plastic to their young. Sadly, many animals die in this way.

The density of plastic also prevents sunlight from reaching algae and plankton below the surface of the ocean. This is a problem because these organisms need sunlight to photosynthesize. The reduced growth of algae and plankton then has knock-on effects on wider marine food webs, all the way up to predator species such as sharks and whales.

No one knows quite how big the Great Pacific Garbage Patch is, but the point is – it shouldn't be there! As the Garbage Patch is in the high seas, no one country has responsibility for it. Even if we could cooperate to clean it up, many problems arise. For example, any nets used to gather the plastic would also gather fish. What we can do, however, is stop it getting any bigger! A number of campaigns and organizations aim to raise awareness of litter in the ocean and encourage its responsible disposal.

HERMIT CRAB LIVING INSIDE A PLASTIC POT.
© Matt Doggett, Earth in Focus

PLENTY MORE FISH IN THE SEA....?

Annie Emery and Nicole Franz, FAO

The ocean, in particular the sunlit zone, provides humans with highly nutritious food, including fish, crustaceans, molluscs, bivalves and aquatic plants (look back at Chapter 2). Around three billion people rely on fish specifically as an important source of protein in their diets (*Source*: FAO, 2012). What's more, around 660-820 million people are employed in the global fishing and fish farming industries (*Source*: FAO, 2012; look back at Chapter 4). For these reasons, it's vital that fisheries are effectively managed to ensure there continues to be plenty of fish in the sea for everyone to enjoy.

The fisheries sector in many developing countries remains small-scale, involving individual fishers or small teams using lines, nets and traps, which are often set by hand. As countries become more developed, their fisheries become more industrial. Many European countries, the USA, and Japan, for example, have large fleets of fishing vessels fitted with hi-tech equipment that fish heavily in their own waters and across the global ocean.

As fishing technology gets more industrial and global demand for fish rises, the impacts of fishing on the marine environment also grow. Particularly popular types of seafood (e.g. some kinds of cod and tuna) have been severely overfished, making it hard for their populations to recover. Some of these species have been put on the IUCN red list of threatened species.

At the same time, it is hard to control exactly which fish are caught in a large net, meaning that fish of the 'wrong' species or size (fish that are not wanted) are often caught as by-catch. Unfortunately, by-catch is often discarded (e.g. thrown back into the sea), which means that certain fish stocks (e.g. plaice and dab in the North Sea) are being

SCHOOL OF FISH.
© Digital Vision/Thinkstock

WOMAN SELLING FISH AT THE MARKET IN BURKINA FASO.
© Alessandra Bendetti, FAO

harmed without actually feeding anyone. Discards can be particularly high from fisheries that trawl the seabed. Not only are overfishing and discards bad for fish species and ocean ecosystems, they also threaten the sustainability of the livelihoods of future generations of fishers.

To manage fisheries more sustainably, many regional fisheries management organizations, national and international governments are imposing fishing quotas: limits on how much of certain kinds of fish may be caught. The European Union sets quotas through its Common Fisheries Policy. This is

currently being reformed and a number of ways to reduce discards are being considered:

:: Helping fishers to develop measures to avoid unwanted by-catch, by using selective fishing gear, or closing certain areas to fishing at important times (e.g. when fish are reproducing).

:: Allowing fishing vessels that are likely to catch

a mixture of species e.g. cod, haddock, and whiting, to have quotas for all these species, rather than just one at a time.

Find out more:
The Seafood Decision Guide: **www.ocean. nationalgeographic.com/ ocean/take-action/seafood- decision-guide**

SCIENCE IN THE SUNLIT ZONE

Scientists study what goes on in the surface ocean in two ways: from space and from research ships.

STUDYING THE OCEAN FROM SPACE

The Earth is under constant surveillance from space. Hundreds of satellites are orbiting the Earth, observing what is going on below. This kind of Earth observation is known as 'remote sensing', and some of these satellites are specially designed to see what is going on above and below the ocean's surface.

An array of different sensors capable of measuring anything from sea surface temperature, to wave height, to ocean colour are attached to these satellites. The colour of the ocean is interesting to scientists as it can be used as a measure of the amount of phytoplankton in the water, and even what kind of phytoplankton are present. This is because different phytoplankton species contain different pigments, which can range in colour from browns to greens

PHYTOPLANKTON BLOOM OFF THE COAST OF SCOTLAND, TAKEN FROM SPACE.
© ESA

to reds. Remote sensing is an incredibly useful tool for marine scientists as it provides up-to-the-minute information on what is happening near the surface of the ocean. (If you remember, remote sensing can also be applied to studying other ocean-related habitats, such as mangroves and salt marshes – refer back to p.116!)

STUDYING THE OCEAN FROM RESEARCH SHIPS

Unsurprisingly, one of the best ways of finding out what is going on in the ocean is to actually go there. Scientists study the ocean from research ships, which are purpose-built to carry out state-of-the-art scientific research. Research ships are designed to have ample laboratory space for large numbers of scientists to take all their equipment on board. They can take the scientists to any ocean in the world, from the steaming tropics to the frozen Arctic.

SCIENTISTS DEPLOYING A CONDUCTIVITY TEMPERATURE DEPTH ROSETTE INTO THE OCEAN, WHICH IS USED TO MEASURE TEMPERATURE, SALINITY, DEPTH AND OTHER THINGS OF INTEREST. IT IS ALSO USED TO SAMPLE WATER FROM DIFFERENT DEPTHS OF THE OCEAN.
© Plymouth Marine Laboratory

CONCLUSION

The sunlit part of the ocean is where most life is found from microscopic bacteria, viruses and plankton to large mammals such as whales, seals and dolphins. This is also where many fishing activities occur. Fishing is essential to the livelihoods of millions of people around the world, but our growing demand for fish is changing marine ecosystems.

Humans are also impacting the sunlit ocean in other ways. The constant motion of the ocean transports life and heat around the ocean, but it also helps transfer people and our rubbish. A vast area of the North Pacific Ocean is now infamous for its slick of plastic waste. No one country has control of the high seas, so getting countries to work together to overcome these problems is challenging.

LEARN MORE

:: Fishwatch: **www.fishwatch.gov**
:: Good Fish Guide: **www.goodfishguide.co.uk**
:: Great Pacific Garbage Patch: **www.greatpacificgarbagepatch.info**
:: Marine Stewardship Council: **www.msc.org**
:: NASA: **http://kids.earth.nasa.gov/seawifs/phytoplankton.htm**
:: Ocean Drifters: **www.www1.plymouth.ac.uk/research/marine/oceandrifters/Pages/default.aspx**
:: Oceanic Research Group: **www.oceanicresearch.org/education/films/planktonfilm.htm**
:: Plastic Oceans: **www.plasticoceans.net**
:: WWF: **www.wwf.panda.org/about_our_earth/blue_planet/problems/problems_fishing**

THE DEEP SEA TREASURE CHEST

THE VAST EXPANSE OF THE DEEP OCEAN IS HOME TO AN ASTOUNDING ARRAY OF BIODIVERSITY AS WELL AS VITAL RESOURCES FOR OUR EVERYDAY LIVES. HOW CAN WE TREASURE BOTH DEEP SEA LIFE AND OCEAN RICHES?

13

David Billett, National Oceanography Centre

Most of the deep sea has never been seen by human eyes and, compared to other marine ecosystems, we know relatively little about it. What we do know shows it to be a fascinating place, containing weird and wonderful marine life as well as amazing seabed features. As many of the species found there are very slow growing they, and the delicate physical structures of the deep sea bed, are extremely vulnerable to disturbance.

THE JABBERWOCKY: A HYDROTHERMAL VENT CHIMNEY 2 800 M BELOW SEA LEVEL, SOUTHWEST INDIAN RIDGE.
© University of Southampton

DEEP SEA **FACTS**

- 88 percent of the ocean is deeper than 1 km.
- 76 percent of the ocean lies at depths between 3 and 6 km.
- Life on land is restricted to a thin veneer generally no higher than the tallest tree, this means that 99.5 percent of the living space on Earth is in the oceans.
- There are about 10 to 15 million different animal <u>species</u> and a billion different forms of <u>microbes</u> in the ocean.
- As one goes deeper into the ocean the mass of the water pressing down from above increases by 1 atmosphere for every 10 m. At 4 000 m this is like an elephant standing tiptoe on the nail of your big toe!
- In most oceans and seas temperature decreases with increasing depth. Waters 4 000 m deep are typically 2 to 4 °C.

BACKGROUND IMAGE
DEEP WATER.
© Barun Patro, SXC

FROM TOP
PILLOW LAVAS ALONG A LARGE FISSURE ON THE GALAPAGOS RIFT.
© NOAA

INACTIVE HYDROTHERMAL CHIMNEYS MADE OF IRON OXIDE ON LOIHI VOLCANO.
© NOAA

MUD SEA FLOORS.
© NOAA

MYSTERIOUS LANDSCAPES

Massive mountain ranges extend through all the major oceans. Measuring about 16 000 km, the Mid-Atlantic Ridge exceeds the length of the Himalayas, Andes and Rockies combined! There are also about 100 000 isolated subsea mountains (seamounts) with towering peaks thousands of metres high. The deep sea also features gorge-like trenches, up to 11 km deep. These form where two massive tectonic plates of the Earth collide, forcing one plate under the other in a process known as subduction. In between the huge climbs and deep drops, vast sprawling flat plains of mud cover about half of the deep ocean floor.

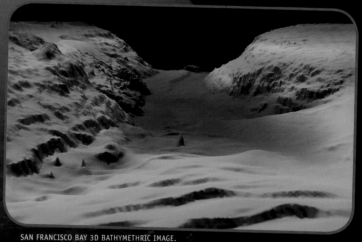

SAN FRANCISCO BAY 3D BATHYMETHRIC IMAGE.
© NOAA

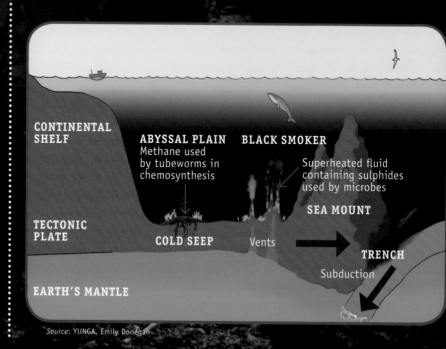

CONTINENTAL SHELF

ABYSSAL PLAIN
Methane used by tubeworms in chemosynthesis

BLACK SMOKER

Superheated fluid containing sulphides used by microbes

SEA MOUNT

TECTONIC PLATE

COLD SEEP

Vents

TRENCH

Subduction

EARTH'S MANTLE

Source: YUNGA, Emily Donegan

DEEP SEA
LIGHT AND LIFE

Deep sea life is adapted to particular combinations of pressure, temperature, light and oxygen, leading to continuous and complex changes in species throughout the ocean.

Sunlight fades rapidly the deeper into the water column you venture. Having said that, in the dark depths of the ocean, the water is ablaze with bioluminescent light made by many organisms. Some of them don't even have eyes to see the light they make. Creepy...

Plants are the essential base of marine food webs, but they cannot survive in the darkness of the deep sea. Instead, new life is formed using the chemical energy of two unique processes. In the first, microbes use methane bubbling up from sub-seabed ice fields of frozen methane to create energy. A second process takes place where superheated hydrothermal vent fluids gush out of the seabed as 'black smokers', allowing microbes to use sulphur compounds, especially hydrogen sulphide, to create organic matter. The peculiar chemistry of hydrothermal vents and cold seeps allow these microbes to support novel life forms of massive tubeworms, giant clams and eyeless shrimps.

The vast majority of animals, however, even in the deepest recesses of the ocean, are dependent on food created by photosynthesis in sunlit surface waters. Different animals have adapted different ways to obtain it:

• Zooplankton balance the competing pressures of finding food and of not being eaten themselves by swimming a 'marathon' each night from darker and safer deep waters into the food-rich surface waters.
• Other species wait for food from the surface waters to sink down into the deep sea. Competition for this food is intense, as there is less and less food the further down you go. Only 1 percent of the organic matter formed at the surface reaches depths of 4 000 m.

'BLACK AND WHITE' HYDROTHERMAL VENT SYSTEM, 2 400 M BELOW SEA LEVEL, EAST SCOTIA RIDGE IN THE SOUTHERN OCEAN.
© University of Southampton

Most of the deep ocean, therefore, is food poor. There are fewer animals compared to the surface waters, with a greater proportion of smaller organisms, and of animals adapted to deep sea conditions, such as sea cucumbers.

Nevertheless, the deep sea zone occupies such a vast volume that it has a profound influence on our global ocean. Healthy deep sea ecosystems are vital for the replenishment of essential nutrients throughout the ocean. These nutrients are fed back to phytoplankton at the sea surface, resulting in the continued production of life-giving oxygen. Without the deep ocean, our lives would grind to a halt.

ANOPLOGASTER
© David Shale

BENTHOCTOPUS
© David Shale

BACKGROUND IMAGE
RING OF FIRE 2006 EXPEDITION. IF WE WERE OBSERVING THIS TYPE OF ERUPTIVE ACTIVITY ON LAND, WE WOULD HAVE TO RUN FOR OUR LIVES! AT BRIMSTONE PIT THE PRESSURE OF 560 METERS (1 837 FEET) OF WATER OVER THE SITE REDUCES THE POWER OF THE EXPLOSIVE BURSTS. ALSO, THE WATER QUICKLY SLOWS DOWN THE ROCKS AND ASH THAT ARE VIOLENTLY THROWN OUT OF THE VENT.
© Submarine ROF 2006, NOAA Vents Program.

PEOPLE
AND THE
DEEP SEA

Humans are changing deep ocean ecosystems directly through fishing, oil and gas extraction and new mining activities. Potential changes also occur indirectly through human influences on the climate and hence on the amount of food available for deep sea organisms.

Fishing: Deep sea bottom **trawling** (the act of dragging a large net across the seabed) occurs down to depths of 1 500 m, and the tops of most **seamounts** within this depth range have already been severely impacted. Some deep sea corals that live on these **seamounts** are more than 2 500 years old, and, once damaged, they can only grow back slowly. Coral impacted by a single trawl in one hour may take hundreds of years to be restored.

Deep sea fisheries generally last only for a few years in any one area before there are no longer enough fish to support continued fishing. This is because the lack of available food means deep sea **species** take much longer to reproduce than their coastal and shelf sea peers.

Oil and gas extraction: Oil and gas activities extend down to 3 000 m. The physical impacts on the seabed are generally small, but whole ocean **basins** can be affected by oil pollution if there is a 'blow out' at some depth, such as the 2010 Deepwater Horizon explosion in the Gulf of Mexico. Deep submarine canyons, some as large as the Grand Canyon, act as 'fast track' pathways for organic **matter** and pollutants from our shores into the abyss.

DECOMISSIONED DEEP SEA TRAWLERS IN GRIMSBY, UK. AFTER THE 'COD WARS' BETWEEN ICELAND AND THE UK IN THE 1970S, MANY BRITISH TRAWLERS WENT OUT OF BUSINESS.
© John Gulliver, Flickr

DEEP WATER LOPHELIA REEF WITH GORGONIANS.
© JAGO-Team GEOMAR Kiel

IMPACTS OF TRAWLING ON DEEP SEA CORALS SUCH AS LOPHELIA.
© Jan Helge Fosså, Institute of Marine Research, Bergen

BIOPROSPECTING

Salvatore Arico, UNESCO

The ocean is unique in its diversity of living organisms, which are amazingly dense in certain parts of the world. In the lndo-Pacific Ocean, for example, there are as many as 1 000 species per square metre. Over millions of years, species have developed unique properties to enable them to cope with extreme living conditions, such as high pressure, temperature and salinity levels. It is the very uniqueness of these properties which offer exciting potential for the development of new drugs and medicines to treat all sorts of human ailments. This search for new products developed from animals and plants is called bioprospecting.

Products based on marine organisms have already found their way onto the market and are now being prescribed to sufferers of asthma, tuberculosis, cancer, Alzheimer's disease, and cystic fibrosis among others. Other industries, such as oil or paper, are also bioprospecting the deep sea with promising results. Marine bioprospecting of the deep seabed is developing rapidly, and there is much interest in the life forms found in and around hydrothermal vents, cold seeps and similar deep sea formations like mud volcanoes and brine pools, as well as on seamounts, which may be home to a particularly high number of endemic species.

Today, there are no legal restrictions on exploring the deep sea for the purposes of research or financial gain when it comes to its living resources. However, this raises a number of questions. Firstly, as this newly discovered 'blue gold' is mostly located in international waters, it can be argued that the genetic resources living in the deep sea belong to humanity as a whole and therefore ought to be exploited equitably. Secondly, if we are to protect these precious resources and the ecosystems in which these are found, we shall have to exploit them in a sustainable manner.

This article was first published under the title The Last Frontier in UNESCO's World of Science Journal (volume 4, issue 2 April-June 2006 p19-23) and is reproduced with the kind permission of the editor. www.unesdoc.unesco.org/images/0014/001453/145347e.pdf

DEEP SEA **MINERALS**

There are three major sources of minerals in the deep ocean: manganese nodules, polymetallic massive sulphides and cobalt-rich crusts.

Manganese nodules are potato-sized, rocky lumps of minerals that lie on the surface of the sediment at depths of 4 to 6 km, mainly in the tropical Pacific and Indian oceans. They grow very slowly - only a few millimetres every million years! They are often called polymetallic nodules because they contain a wide variety of metals, such as nickel, copper and cobalt. Nodules with the greatest value lie across a vast area between Hawaii and Mexico in the Pacific Ocean. It is estimated that there are 500 billion tonnes of nodules that could be mined in the deep sea...

Cobalt-rich crusts slowly build up layers about 25 cm thick on the sides of seamounts, generally at depths of 800 to 2 500 m. They have taken up to 65 million years to form. The richest crusts are found in the tropical western deep Pacific Ocean.

Polymetallic massive sulphides are formed from superheated fluids seeping out of hydrothermal vents in the sea floor as they cool at or near the seabed. Hydrothermal vent chimneys are made of these sulphides. Some sulphide deposits are ten times richer in metals, such as gold and silver, than land deposits.

MANGANESE NODULES.
© Koelle, WMC

COBALT.
© Jurii, WMC

GROTTO HYDROTHERMAL VENT, ENDEAVOUR RIDGE RIFT VALLEY. THE BLURRY FLUID
COMING OUT OF THE VENT IS 300 °C TO 400 °C HOT AND CONTAINS HIGH CONCENTRATIONS
OF SULPHIDES, WHICH OVER TIME HAVE FORMED THIS HUGE CHIMNEY.
© Neptune Canada, Flickr

MANAGING DEEP SEA
MINING

Mining in the deep sea is remote from where we live, which has the advantage of being out of sight and "not in our back yard". But that doesn't mean we should ignore their impact! These minerals are found in places that we know little about and scientists unfortunately often don't yet understand the effect that mining might have on deep-sea ecosystems.

Many mineral deposits occur in international waters and are the common heritage of all nations. The United Nations has set up a special body, the International Seabed Authority, based in Kingston, Jamaica, to manage mining activities in international waters, and to divide up any profits from mining to individual States, including developing nations.

There are a number of formidable impacts that need to be overcome if deep sea minerals are to be exploited at a large scale:

• The physical impact of a mine, which could involve removing the side of a mountain or a field of nodules. Nodules are a hard surface in an otherwise muddy area and so act as habitat for animals such as sponges, anemones and soft corals. Once a nodule is removed, it will take millions of years for new nodules to form.

• Mining operations will disturb the seabed and create a plume of sediment which will disperse tens of kilometres from the mine site, smothering animals on the seabed.

• A mix of sediment with nodules, crusts or sulphides will be brought to the surface ship for initial processing. The waste produced (known as tailings) will then be released back into the sea.

• Chemical changes may introduce toxic pollutants into the marine environment.

Many of the minerals of interest to deep sea mining are used in your mobile phone, PC and television. It will be up to you to decide what level of impact on deep sea ecosystems is acceptable so that you can continue to use modern technologies!

OPPOSITE PAGE
PALMTREE-LIKE CRINOIDS, ANIMALS THAT CAN MOVE SLOWLY ABOUT THE SEA FLOOR.
© NOAA

Few of us know much about how the metals we use in our homes and in our pockets every day are produced, what effects their mining has on <u>ecosystems</u> and how we can plan to reduce environmental impacts to a minimum.

Take your mobile phone, for example. There is more to it than just the plastic casing; it has metals in the battery, the screen, the chip and in the circuitry. Mobile phones may contain silver, gold, lithium, zinc, nickel, copper and manganese as well as other metals with unique properties.

At present, most of these metals come from land-based sources, but industry is looking for alternative sources, including the deep sea. This is because metal prices are rising due to ever-increasing demand and because some resources can currently only be obtained from one or two countries, which can then control their market value, pushing prices up further still.

SCIENCE IN THE DEEP SEA

Many scientists are trying to understand what lives in the deep sea and how it survives, but studying the deep oceans is extremely challenging: the pressure is too great for people to dive to depths of more than about 150 m. Instead, deep sea researchers use underwater vehicles. Some of these can carry people, such as the Deepsea Challenger, but many are robotic. These vehicles carry a range of equipment to collect images and specimens; some are pre-programmed independent systems, while others are controlled by people on the deck of a ship above.

THE DEEPSEA CHALLENGER

Challenger Deep is the deepest known point on Earth, and is located in the Marianas Trench in the Western Pacific Ocean. At 11 km deep, it's deeper than Mount Everest is tall! In March 2012, a team of scientists and engineers led by film director James Cameron sent a specially designed vehicle – the Deepsea Challenger – into Challenger Deep to collect information about this unexplored and fascinating environment. The Deepsea Challenger was fitted with cameras and a robotic arm to collect samples.

Fun facts:

- It took the submarine about two hours to reach the bottom of the Challenger Deep, descending at about 150 m per minute.
- The water vapour produced when the pilot of the Deepsea Challenger breathed was collected so that, in an emergency, he could drink it!
- The pilot's room – a sphere just 109 cm wide – was so small he had to sit with his legs bent for the whole journey!
- The pressure at the bottom of the Marianas Trench is about one thousand times greater than on the surface – without the pressure control in the Deepsea Challenger and other deep sea vehicles, exploration would be impossible for humans!

THE DEEPSEA CHALLENGER - NOTE THE WARNING TO OTHER MARINE VEHICLES THAT THIS ISN'T JUST AN UNMANNED PROBE BUT THAT THERE'S SOMEONE INSIDE!
© Ben R, Flickr

CONCLUSION

Dark, cold and mysterious, the deep sea is a difficult ecosystem to explore. The animals living there have developed many ways to survive in this extreme environment, often by living longer and reproducing at a lower rate than similar animals in shallow waters. Some can produce their own light, while unique life forms may be found at vents and seeps where microbes sustain food webs through chemical reactions.

We know relatively little about the deep sea, but we are beginning to realise just how important an ecosystem it is. Not only does it support life in other parts of the ocean, but it is also home to natural resources of interest to humans, including novel genes and minerals such as nickel, copper, cobalt, gold and silver. Many minerals are yet to be exploited, but human activities, such as deep sea bottom trawling, are already having major impacts on deep-water corals. Deep sea communities are also threatened by the results of other human activities, such as seabed blowouts which release large quantities of oil (as in the Gulf of Mexico in 2010). It is time to start thinking about what level of human impact is acceptable on deep sea ecosystems in pursuit of what we need for our everyday lives.

LEARN MORE

:: Deep Sea Challenge: **www.deepseachallenge.com**

:: James Cameron's first footage from the deep sea floor:
   www.youtube.com/watch?v=FGzaUiutuRk

:: MarineBio Conservation Society: **http://marinebio.org/Oceans/Deep**

:: NeMO Explorer: **www.ocean.si.edu/deep-sea**

:: The Smithsonian: **www.ocean.si.edu/deep-sea**

:: Superyacht submersibles: **www.scubacraft.com/news/articles/showboats.pdf**

Section

TAKING ACTION FOR THE OCEAN

Chapter 14
THE OCEAN AND YOU!

THE OCEAN AND YOU!

WHY DON'T YOU START A PROJECT TO HELP PROTECT THE OCEAN?

14

Jennifer Corriero, TakingITGLOBAL
Alashiya Gordes, FAO

Having read this Guide, you now know how essential the ocean is to life on Earth. So now it's time to take action to help protect the ocean! Think about the threats the ocean faces – which is the most important to you? Young people around the world are already leading successful projects to help protect and conserve the ocean environment and marine biodiversity. Now it's your turn to take action: read on to learn the six simple steps to start an ocean action project!

DIVER EXPLORING THE WONDERS OF THE OCEAN.
© Comstock

BE SAFE AND SOUND

You can start an ocean action project wherever you live. However, if your project involves you going to the seaside, make sure you take precautions for your own safety, as the ocean can be an unpredictable place. You should also make sure that you don't hurt the marine environment. Remember: "take only pictures and leave only footprints". Please consider the general precautions below and carefully evaluate which other safety issues need to be taken into consideration before you get started.

The Marine Life Information Network for Britain and Ireland has put together a seashore code containing advice on how to look after yourself and to protect the animals and plants that live on the shore (**www.marlin.ac.uk/pdf/seashorecode.pdf**). It says:

:: **Before you go, tell someone where you are going, when you will be back and make sure you know what the weather and tides will be like. If you can, take a mobile phone.**

:: **Walk carefully over rocks, they may be slippery or unstable. Cliffs should also be avoided as they may be unstable.**

:: **Do not take living plants or animals home with you. If you do take shells home, make sure they are empty.**

:: **Take your litter home, it can be dangerous to people and wildlife and can ruin the scenery you are there to enjoy.**

:: **Report anything unusual that you find, but do not touch anything unless you are sure it is safe.**

:: **Treat all living things with respect and replace any stone or seaweed exactly where you found it.**

:: **Wash your hands before you eat and when you get home!**

A FEW ADDITIONAL TIPS SHOULD ALSO HELP TO KEEP YOU SAFE:

:: Avoid muddy shores as you can easily get stuck in the mud.

:: Watch out for waves, especially near rocks, as they can be bigger and more powerful than you think.

:: If there are any warning signs on the beach or coast (such as beach closed or no swimming), make sure you follow the advice.

:: If you want to swim, don't go into the water unsupervised and, if possible, only swim at beaches where there is a lifeguard on patrol. Make sure you know where other people in your group are.

:: Don't swim immediately after a meal.

:: Don't swim near pipes, outflows, rocks, breakwaters and piers and don't use them to jump off.

:: If you get into trouble in the water, don't panic; raise one arm up and float until help arrives. If you find you are in a rip current or undertow, float with it; don't try to swim against it.

:: Only use a snorkel if you are a good swimmer and the water is calm.

:: Take a first aid kit to the beach with you, just in case.

:: If you decide to upload pictures or videos to the internet on Web sites such as YouTube, always make sure that everyone in the pictures or video (and/or their parents) have given their permission before you post anything online.

SIX SIMPLE STEPS
TOWARDS CHANGE

These *Six Simple Steps Towards Change* have been adapted from the Guide to Action created by TakingItGlobal, in consultation with young leaders around the world.

You can use these steps to help you plan and execute your own ocean project:

1. REFLECT AND GET INSPIRED
2. IDENTIFY AND GET INFORMED
3. LEAD AND GET OTHERS INVOLVED
4. GET CONNECTED
5. PLAN AND GET MOVING
6. HAVE A LASTING IMPACT

ICEBERG IN GREENLAND.
© Brocken Inaglory, WMC

REFLECT & GET INSPIRED

Think about the changes you would like to see, whether they are in yourself, your school, your community, your country or even across the whole world. Also think about who or what inspires you to take action. Seeking out sources of inspiration can give you great ideas and help you find the strength to turn your vision into a reality.

1

HAVE A LASTING IMPACT

Monitoring and evaluation are important parts of project management. Throughout your project, you'll want to identify the obstacles you are facing and the lessons you are learning. Remember, even if you don't achieve all of your expectations, you most probably influenced others and experienced personal growth! At the end of your project, you can revisit your notes and think about how you can benefit from this experience in your NEXT project... Encouraging other youth to get involved in the issues you care about is also a great way to sustain your efforts, even after your own project has ended.

6

PLAN & GET MOVING

Now that you are equipped to take action, it's time to begin the serious planning... You already have an idea of the issue or issues you'd like to work on: now identify a particular goal you can work towards. When you have your plan, stay positive and focused. If you encounter obstacles, don't worry, that's completely normal! You will learn a great deal from overcoming challenges.

5

IDENTIFY & GET INFORMED

Which issues are you most passionate about? Learn more about them by gathering information about the things that interest you. By informing yourself, you are preparing yourself to tackle the challenges that lie ahead.

LEAD & GET OTHERS INVOLVED

Being a good leader is about building on the skills you have and knowing how to leverage the strengths of others. Write down the skills that you and your team members bring to your action project and see how each of you can use your strengths to lead in different ways. Remember that good leadership includes great teamwork!

GET CONNECTED

Networking can give you ideas, access to knowledge, experience and help in gaining support for your project. What are you waiting for? Create a map of your networks and start tracking your contacts!

STEPS FOR CREATING AN OCEAN PROJECT.
Source: Adapted from *Guide to Action: Simple Steps Towards Change*, TakingITGlobal, 2006

1. REFLECT AND GET INSPIRED
Reflect on the issues you are passionate about

Take a moment to **REFLECT** on the threats facing the ocean that matter most to you. Imagine how it would be if we humans lived in harmony with our beautiful blue planet and its natural systems. *What would that world look like?*

Think about which threats to the ocean concern you the most. Or think about which ocean plants or animal species, habitats or ecosystems you want to conserve, protect and restore locally and globally.

CONSERVE – Conservation means to preserve the natural functions of ocean ecosystems and biological marine communities, as well as their resilience (their ability to recover from shocks). This can be done by limiting the use and extraction of the natural resources they provide.

PROTECT – You can help protect an ecosystem or species by campaigning to have it protected by your government's laws or international policies.

RESTORE – Restoration helps 'repair' degraded ecosystems or habitats to a more natural, less damaged state so they can function better again.

Get inspired

GET INSPIRED by learning about local and international ocean champions – reading the case studies of youth-led ocean projects in this chapter is a good place to start! You can also start to identify local ocean champions in your family, neighbourhood, school or city.

Join TakingITGlobal's network of youth engaged in global issues and connect with youth leaders, organizations and projects from all over the world at: www.takingitglobal.org

ASK YOURSELF

Are there threatened marine plant or animal species you want to protect?
Are there certain ocean habitats or ecosystems you would like to conserve or restore?
Which threats to the ocean worry you most?
Do you know anyone who is being affected by threats to the ocean?
How about communities in other countries?

CASE STUDY: **LUISA SETTE CAMARA**
21, Abrolhos, Brazil

As part of my university studies in law, I researched marine underwater noise pollution and its impacts on marine mammals in Abrolhos Marine National Park. Since conducting this research and understanding the threats Abrolhos faces, I have been advocating for better protection of the site.

Abrolhos inspired me deeply. It was incredible to experience how strong the relationship is between the Abrolhos' local community and the ocean. The community respects nature and recognizes its intrinsic value. There is no need for research or economic valuations to show them the importance of the volcanic archipelago - they already know it. This relationship is reflected in the local culture, which involves a folkloric celebration called 'Marujada'. Songs related to the sea are important elements of the celebration. From a natural perspective, Abrolhos is a unique site that is a home and refuge for thousands of species. A wide range of threatened species still live there – for example, it is a breeding area for humpback whales, which have almost been driven to extinction by commercial whaling.

I see Abrolhos as a place synonymous with life; a place in which to find peace from the chaotic outside world – for all living organisms, including mankind.

A HUMPBACK WHALE DIVES IN ABROLHOS.
© Jonathan Wilkins, WMC

2. IDENTIFY AND GET INFORMED
Identify the issues you are going to take action on

Referring to your reflections on the marine threats that you are most concerned about or the <u>marine resources</u> that you would like to conserve, protect or restore, **IDENTIFY** and narrow down the issues that are most important to you.

Develop a set of question that you want to answer. Here are some ideas:

- What makes this issue unique and important?
- Who is most affected by the issue, and why?
- How does this issue differ locally, nationally, regionally and globally?
- What different approaches have been taken to understand and tackle the issue?
- Which groups are currently working on addressing the issue? (Consider different sectors such as governments, corporations, non-profit organizations, youth groups, United Nations agencies, etc.)

Get informed

GET INFORMED by finding resources on the issues you are interested in. Be sure to check out the links listed at the end of each section of this Guide! You can also visit TakingITGlobal's Issues pages to find further organizations, online resources and publications for inspiration:

www.tigweb.org/understand/issues

Make a list of all the key resources you have found (organizations, publications, Web sites):

1.........................

2.........................

3.........................

4.........................

ASK YOURSELF

What more can I learn about the issues I care about?

CASE STUDY: SAVING THE ARCTIC

LOOKING AT POLAR BEARS.
© WAGGGS

In 2012, two young women from the World Association of Girl Guides and Girl Scouts (WAGGGS), Sena Blankson from Ghana (19) and Miryam Justo from Peru (30) joined Greenpeace's expedition in support of the Save the Arctic campaign. The voyage's mission was to measure ice thickness, density and solidity using a variety of techniques and tools to assess how climate change has been affecting the Arctic. It also aimed to allow people from each of the world's continents to bear witness to what's happening to the Arctic and its surrounding ocean due to climate change. Sena and Miryam observed, assisted and learned from the scientists on the expedition. Equipped with their new knowledge and moved by their experiences, they are now working hard to spread the word to the world: Arctic ice cover is decreasing – 75 percent of it has already melted! As the Arctic's natural cooling effect on Earth's climate diminishes along with its ice cover, the whole planet will feel the results (e.g. in the form of warmer average temperatures and changed rainfall patterns). Knowing all this, we also know that protecting the Arctic needs to be a priority for us (look back at Chapter 11) - starting today!

Find out more:

:: **www.savethearctic.org**
:: **www.wagggs.org/ en/flagforthefuture/ girlsgotothearctic**

ABANDONED COALMINE IN SVALBARD.
© WAGGGS

GETTING AROUND.
© WAGGGS

3. LEAD AND GET OTHERS INVOLVED
Lead your project to success

Identifying your skills and characteristics will help you **LEAD** your project to success. Start by understanding your own strengths and needs, and then consider how creating a team could help you to better achieve your goals. Helping your team members identify and leverage their own strengths and talents for the project is an important part of leadership. It is also important to ensure that all those involved are able to share in the vision of what you are trying to achieve. Think of someone who shows strong leadership. What makes that person a good leader? Create a list of leadership qualities. Some examples are:

- Accountability
- Compassion
- Dedication
- Fairness
- Honesty

- Innovativeness
- Being motivational
- Open-mindedness
- Responsiveness
- Being visionary

Build a team and get others involved

Once you have reflected on your personal leadership assets and goals, you are ready to develop a team and **GET OTHERS INVOLVED**. Start with people you know, and then expand the project to people they know and so on – your team will grow quite quickly! When you feel ready, you can call for participation from the wider community, too. How can you encourage them to take part in your project to address the ocean issues that you are most concerned about?

List the leadership skills that you possess:

1 ....
2.........................
3.........................
4.........................

List the leadership skills that you want to develop:

1.........................
2.........................
3.........................
4.........................

Name some people you already know who would want to be part of your team:

1.........................
2.........................
3.........................
4.........................

What are some of the skills that your team members can contribute?

1.........................
2.........................
3.........................
4.........................

CASE STUDY: **MARLON WILLIAMS**
28, Belize Barrier Reef, Belize

MARLON CAPTURING INVASIVE LIONFISH.
© Toledo Institute for Development and Environment (TIDE)

I was raised in a 'tough' fishing family on the Rio Grande River in Southern Belize. Many of our fishing practices (e.g. loglines and baiting, nets and harpooning) were passed down from my grandparents. Many of these practices were actually illegal, but as they were highly effective and we had to make a living, we used them anyway. To this day, I remember how all of them work. However, as I grew older, I also started learning about how fish stocks are declining all over the world, and that the practices my family and I were using were a small part of a big problem. I understood that things had to change, and saw the opportunity for my family and my community to find a more sustainable way of living. Unfortunately, many people still believe that fish stocks will never run out, and this mindset urgently needs to change. I don't expect every fisher in my community or Belize to change the way they fish immediately, but I can try to make sure they understand the very basics. If people understand that fishing without allowing fish stocks to recover will mean that one day, they, or their children, will have no fish left to catch, they will eventually also understand the broader concepts of resource protection and nature conservation. When I make tours around the Belize Barrier Reef where I work now, I share my story, telling people about my change from employing many illegal fishing practices to becoming a conservationist and marine biologist. I believe that by being open and honest, we can touch people's lives and change the world.

MEASURING LIONFISH.
© TIDE

RELOCATING A TURTLE NEST TO KEEP IT SAFE.
© TIDE

4. GET CONNECTED

You can also develop a team by networking and **GETTING CONNECTED** to people you have not yet met, but would like to work with. They might be associated with people whom you already know; alternatively, why not connect with a network that works on issues that matter to you?

You can start by attending events and conferences about ocean conservation. Do a little bit of research and find out what opportunities there are in your area.

List at least one event in your area that you would like to attend:

....
....
....
....
....
....
....
....

OCEAN INITIATIVES: BRINGING YOUNG ACTIVISTS TOGETHER

There are many initiatives for you to join or get inspired by! Here are just a few examples:

:: Starting in 2007 as part of the EU's work on sustainable maritime policies, The World Ocean Network and its partners organized a series of Youth Parliament meetings around the world.

Find out more here: **www.worldoceannetwork. org/EN/page-ACT_ WITH_US-Youth_ Parliament_-5-108.htm** and read the highlights of the Youth Declarations for the Ocean produced during these consultations here: **www.unesco.org/ science/doc/ioc/ YouthDeclaration.pdf**

:: The East Asian Seas Youth Forum: Young Champions for the Oceans met for the third time in 2012 in the Republic of Korea. Take a look at the presentations given and the Declaration formulated during the Forum here: **www.eascongress.pemsea. org/yf3**

:: The MOTE Marine Laboratory in Florida, USA holds an annual Youth Ocean Conservation Summit. Read up here: **www.stowitdontthrowitproject.org/pb/wp_a9baf081/wp_a9baf081.html**

:: The International Youth Forum Go4BioDiv lets young people engage in the Conferences of the Parties of the Convention on Biodiversity (CBD-COP).

Watch the Go4BioDiv documentary about Marine and Coastal Life here: **www.go4biodiv.org/news/2013/03/07/go4biodiv-documentary-ready**

:: As part of the thinking process going into formulating new development goals to follow the Millennium Development Goals after 2015, the Global Ocean Commission and partners are mobilizing support for a Sustainable Development Goal dedicated to protecting the ocean. Find out more: **www.globaloceancommission.org/policies/a-sustainable-development-goal-for-the-global-ocean**

What other initiatives can you find? Which ones would you like to get involved in?

BIRDS AND BEACH, PERSIAN GULF COAST OF SAUDI ARABIA.
© Don Toofee, WMC

5. PLAN AND GET MOVING

Develop an action plan

By now, you have identified key threats to the ocean, you've learned more about them and have a good idea of your skills as well as those of your team. You have also learned about the importance of networking and connecting with people who can help you to achieve your goals. This means you are now ready to develop and implement your action **PLAN**!

Keeping in mind the main issue you have identified, which goal or desirable outcome would you like to work towards in your action plan? Here are some possible examples to get you thinking:

Conserve
- Campaign to prevent the pollution of a particular marine <u>habitat</u>.
- Raise awareness about a product, leisure activity or industrial activity that threatens the ocean environment.

Protect
- Campaign to have an <u>ecosystem</u> recognized as a United Nations Educational Scientific and Cultural Organization (UNESCO) Biosphere Reserve.
- Advocate for an at-risk underwater plant or animal <u>species</u> to be included in the International Union for the Conservation of Nature (IUCN) Red List of Threatened Species.

Restore
- Organize or participate in a beach clean-up to restore a local coastal environment.
- Launch a campaign advocating sustainable seafood consumption in your community to help threatened local fish <u>stocks</u> to recover.

Develop a mission statement

Project mission
Clarify what you want your project to achieve and write it down in the form of a mission statement: a short, clear sentence about your purpose. For example: *Restore undisturbed turtle-nesting on local beaches.*

Write your goals:

Brainstorm five possible actions related to the ocean issue you have identified. Actions are activities that will help you achieve your goals:

1 ....
2 ....
3 ....
4 ....
5 ....

Project activities

What actions can you take to work towards achieving your project mission? For example: *Run an awareness campaign about turtle-friendly recreational beach activities.*

Break it down

You know your mission. Now, use the sample chart below to break your project down into specific activities, resources, responsibilities and deadlines. Planning these activities in detail will ensure your project is a success. If your goal is to *restore undisturbed turtle-nesting on local beaches*, your chart might look similar to this example:

ACTIVITY	RESOURCES	RESPONSIBILITIES	DEADLINE
Restore undisturbed turtle-nesting on local beaches.	:: Local conservation groups :: Local city council :: Keen friends and family members :: ...and many more!	:: *I will:* consult local conservation groups and the city council about how best we can make sure our activities on the beach don't disturb nesting turtles. :: *Mum will:* help me write a newspaper article about turtle-friendly beach activities for our local newspaper. :: *John and Jane will:* design posters and leaflets for our campaign. :: ...and so on!	8th June, World Oceans Day

Implement

Once you have finalized your plan, it's time to GET MOVING and actually implement your project! Take time to chart your progress so that you can appreciate and assess the impact of your actions. Document your project with pictures and videos. It is also a good idea to keep a project journal or blog!

Try to refer to your plan along the way, but don't be surprised if not everything goes according to it. Circumstances may be unpredictable, so it's important that you are flexible while staying organized: you may need to revise your plan as you encounter challenges. So, remember to enjoy the entire experience as a learning process.

SECTION D :: TAKING ACTION FOR THE OCEAN / CHAPTER 14 :: The ocean and you!

Raise awareness

Create promotional materials, such as press releases and flyers, to get publicity and to let people know about your project! Word of mouth is one of the strongest marketing tools. Do you use social media like Twitter, Facebook or Youtube? Share your passion for the ocean with your friends and followers! Be enthusiastic and stay positive when you let others know how and why they should get involved. Another way to promote your project is to create a project page on TakingITGlobal (**www.takingitglobal.org**) or add it to The Green Wave website (**www.greenwave.cbd.int**).

Stay motivated

Be sure to stay motivated, especially if you find yourself facing obstacles. Remember: every challenge is an opportunity to learn. Use your creativity to come up with innovative solutions to each challenge: that's problem-solving in action!

CASE STUDY: **BOYAN SLAT**
19, The Ocean Cleanup Array

In 2012, Boyan Slat, a Dutch engineering student, unveiled his prize-winning idea to clean up the plastic rubbish adrift in the five ocean gyres. Today, he is working with a team of 50 engineers to hone his design and finalize the project's feasibility study. The basic concept is to use long floating booms that move with the ocean's natural currents to channel the waste into fixed plastic processing stations. The gathered plastic could then be recycled.

The idea is very exciting, but of course careful planning and evaluation is necessary before a project as big as this can be implemented. Visit Boyan's website to find out about the problems the team is trying to work around and to take a look at their fundraising drive to get an idea of what it's like

BOYAN PRESENTING HIS IDEA IN THE NETHERLANDS.
© Boyan Slat

to gear up for a project on a large scale:
www.boyanslat.com/plastic5

Watch Boyan talk about his project here:
www.boyanslat.com/TEDx.

6. HAVE A LASTING IMPACT

Monitoring your project throughout each stage will help you to respond appropriately to changes that occur along the way and **HAVE A LASTING IMPACT**. It is helpful to set out indicators or measures of success to make sure you stay on track. The more specific your indicator, the easier it will be to evaluate your achievements. For example:

OBJECTIVE	INDICATORS OF SUCCESS
Restore undisturbed turtle-nesting on local beaches.	:: Number of people (and/or communities) engaged in project.
	:: Number of beaches targeted by project.
	:: Number of materials created and distributed as part of the project.
	:: Decrease in turtle disturbing activities after the campaign (try to quantify).
	:: Number of turtle nests on local beaches a little time after the main campaign. (Make sure a conservationist shows you how to count these carefully without disturbing the turtles!)

CASE STUDY: **EMILIE NOVACZEK**
24, Coral restoration in San Andrés, Colombia

We often call coral <u>reefs</u> the rainforests of the sea, because they're a hotspot for marine life. It's estimated that <u>reefs</u> support almost a third of all ocean <u>species</u>, even though they cover less than 1 percent of the ocean floor. Unfortunately, these incredibly <u>biodiverse</u> <u>ecosystems</u> are vanishing. Experts predict that one-third of all <u>reef</u>-building corals are at risk of immediate <u>extinction</u>.

↘

EMILIE'S COLLEAGUE MONITORING CORAL.
© Emilie Novaczek

In 2012, I was working with the Seaflower Marine Protected Area in San Andrés, Colombia. My favourite task was supporting a low-tech coral restoration project. Coral restoration is a relatively new field, and in the Caribbean we focused on two critically endangered kinds of coral: staghorn coral (scientific name: Acropora cervicornis) and elkhorn coral (scientific name: Acropora palmata).

THE CORAL NURSERY.
© Emilie Novaczek

These corals build great reefs, which act as habitats for many other species. However, they're particularly vulnerable to disease, water temperature-related bleaching, and damage by storms or boats. Worryingly, staghorn and elkhorn coral populations have declined by 80–90 percent throughout the Caribbean and western Atlantic in less than 30 years.

Coral restoration techniques were designed to reinforce the remaining populations. In San Andrés, we collect wild staghorn fragments broken by boat hulls that scrape against the reef. We carefully attach them to an underwater rope structure, which we call 'the nursery' (because we raise and nurture them there). Scientists, like me, visit regularly to measure growth, remove algae or predators and treat diseased fragments.

The coral fragments grow quickly, and when they reach a healthy size (about 10 cm) we plant them on shallow reefs where they used to grow. We dive down to scrub algae off surfaces (e.g. underwater boulders) were corals used to grow, hammer a long nail into the surface and fix the young coral firmly to the nail with plastic zip ties so it has the support it needs.

Last year I planted almost 70 staghorn fragments. My hope is that my colonies will be building reefs and, as a result, important natural habitats, for hundreds of years. But of course, this will only work if human activity, pollution and climate change do not make it impossible for coral to survive.

CONCLUSION

Now that you have read through the *Six Simple Steps Towards Change*, you are ready to lead your own ocean action project to success! Remember that these steps are only guidelines and you may want to set your own path. There is no perfect recipe for success because each situation is unique. Every action project you start is a learning process that will challenge you to solve problems and develop your own skills and talents.

Don't forget to take the time to document and reflect on your progress. Keeping good records will help you learn from your experience and will make it easier to share what you have learned with other people at home and abroad. As a young ocean champion, you can help other youth to reflect, get inspired and start their own action projects!

Use the activity-packed
Ocean Challenge Badge
to inspire you to take action!
www.fao.org/docrep/018/i3465e/i3465e.pdf

LEARN MORE

:: The Green Wave: **www.greenwave.cbd.int**

:: *Guide to Action: Simple Steps Towards Change,* TakingITGlobal (2006): **www.tigweb.org/action-tools/guide/online.html**

:: Ocean Today: **www.oceantoday.noaa.gov**

:: TakingITGlobal: **www.takingITglobal.org**

:: TakingItGlobal for Educators (TIGed): **www.tigweb.org/tiged**

:: WAGGGS: **www.wagggs.org**

:: World Oceans Day: **www.worldoceansday.org**

:: WOSM: **www.scout.org**

AQUARIUM
IN AN ANNEX!

ANNEX

A

**IMMERSE YOURSELF IN OUR AQUARIUM
IN AN ANNEX!**

Emily Donegan and Alashiya Gordes, FAO

**Whether you're an aquarium veteran or a first-timer, we're pretty sure you haven't
heard of most of the cool and crazy creatures we have in store for you...**

The ocean is full of amazing forms of life, and many of these are catalogued in the World Register of Marine Species (WoRMS: www.marinespecies.org). WoRMS contributed to The Census for Marine Life (www.coml.org), a ten-year international project that ended in 2010 – over 2 700 scientists in more than 80 countries participated. The Census set out to assess how many different kinds of marine life exist, where these organisms live and how many of each exist (their 'abundance'). The Census successfully described 1 200 new species and increased the estimate of the number of species living in the ocean from 230 000 to nearly 250 000 (and that still excludes microbes!). This research is extremely useful, as it gives us a better idea of the marine life that exists today, allows us to measure how numbers are changing and helps us make informed estimates about how they may keep changing in the future. So read on for a little taste of some of the weird and wonderful life in our ocean...

WELCOME TO
THE AQUARIUM

→

ANIMAL CHAMPIONS

COLOSSAL SQUID, ESTIMATED TO GROW UP TO 14 M IN LENGTH, HERE COMPARED TO THE SIZE OF A HUMAN.

AN ARCTIC TERN IN FLIGHT.
© Andreas Trepte

THE FABLED KRAKEN

The **COLOSSAL SQUID** is massive. It's the world's largest <u>invertebrate</u>, and can grow up to 14 m long. As they live in the deep freezing Southern Ocean, colossal squids have rarely been seen alive, so not much is known about them. However, we do know that their brains are pretty small – and doughnut-shaped! The oddest thing about this is that their oesophagus runs through the 'doughnut hole', meaning they have to chew up their food really well to prevent food lumps from giving them brain damage when they swallow! We also know that they are preyed on by sperm whales – and that sperm whales often have nasty scars and circular cuts from the hooks and suckers of the squid's tentacles. The colossal squid and its slightly smaller cousin, the giant squid, have inspired myths and legends in sea-going cultures all around the world...

FLYING TO THE MOON

The little white **ARCTIC TERN** is by far the most hardcore of animal travelers! It <u>migrates</u> from the North to South Pole each year – a distance of about 70 000 km! It lives for approximately 30 years, so it could tot up over two million kilometers during its lifetime. That's enough to go to the Moon and back three times!

MANTIS SHRIMP SEEN
IN ALOR, INDONESIA.
© prilfish

THE FASTEST TRIGGER IN THE OCEAN

MANTIS SHRIMPS are thought to have the most complex eyes in the animal kingdom. We humans can detect three colours (red, yellow and blue). All the other colours we see are derived by mixing these together (green, purple, brown, etc.). A mantis shrimp can detect at least eight colours – that's five colours we can't even imagine!

GIANT OARFISH FOUND ON THE PACIFIC OCEAN
SHORE NEAR SAN DIEGO, CALIFORNIA.
© LT DeeDee Van Wormer

Mantis shrimps also make one of the fastest movements in the animal kingdom: they have a claw that they use to break open the shells of their prey, which can accelerate to the speed of a bullet at 23 m per second. This is so fast that the water boils around the claw as it strikes! If a human were able to accelerate their arms at just a tenth of that speed, they could throw a baseball into orbit!

THE KING OF HERRINGS

The **GIANT OARFISH** is the longest bony fish, reaching lengths of up to 11 m. It is silver, long and ribbon-like and has a red crest on its head that looks a bit like a crown, hence its nickname: King of Herrings. It is believed that the King of Herrings is responsible for many of the reports of sea-serpents...

RARE AND ENDANGERED

THE SPORTS-CAR OF THE SEA

BLUEFIN TUNA can grow up to 4 m long, making them one of the biggest bony fish out there. Their gills need a constant flow of water over them in order to get oxygen, so tuna can never stop swimming, or they would suffocate. And tuna swim fast – very fast. They can reach speeds of between 70 and 100 km per hour, after accelerating faster than a sports car – and they can cost more than a sports car too! Their meat is a prized delicacy; in January 2013, a 222 kg bluefin was sold in Japan for 155 million Yen (that's US$ 1.7 million)! Unfortunately, **overfishing** has left these big, fast fish critically endangered. The old saying 'there are plenty more fish in the sea' sadly isn't true for the tuna.

GROUP OF ATLANTIC BLUEFIN TUNA IN FAVIGNANA, SICILY, ITALY.
© Danilo Cedrone, FAO

BACK FROM THE DEAD?

The **COELACANTH** was thought to have become **extinct** 65 million years ago around the same time as dinosaurs. Then, in 1938, a museum curator recognized a very interesting fish indeed in the catch of a local fisherman in South Africa. It was a coelacanth, alive! Coelacanths are actually more closely related to mammals like us humans than they are to ray-finned fish (like tuna, tilapia or your pet goldfish). They are one of the rarest animals in the world.

LEATHERBACK TURTLE NESTING NEAR GALIBI, SURINAME.
© JuliasTravels

THE RECORD-BREAKING LEATHERBACK

The **LEATHERBACK TURTLE** is a pretty unique animal. It is the largest turtle and the only one not to have a bony shell (hence the name 'leatherback'). It is also the fastest moving reptile (swimming at up to 35 km per hour - and people normally think of turtles as slow!), one of the deepest diving marine animals (down to almost 1 300 m!) and has set the record for the longest **migration** through the sea of any vertebrate (over 20 000 km), before the signal from the turtle's radio collar was lost! However, these huge turtles are being put in danger by carelessly discarded rubbish. For example, floating plastic bags frequently suffocate the turtles because they look very much like a tasty jellyfish to a hungry leatherback.

PRESERVED COELACANTH SPECIMEN IN THE NATURAL HISTORY MUSEUM, VIENNA, AUSTRIA.
© Alberto Fernandez Fernandez

WEIRD AND WONDERFUL

GOOSE BARNACLE FROM WATTAMOLLA
BEACH, SYDNEY, AUSTRALIA.
© Schomynv

THE BARNACLE GOOSE AND THE GOOSE BARNACLE...

The barnacle goose is black, grey and white. The **GOOSE BARNACLE** is also black, grey and white and supposedly shaped a bit like a goose's body... This vague similarity prompted medieval authors to affirm that these barnacles were the goose's **eggs** which grew on floating wood in the sea, and flew off as feathered geese when they were ready – in time with their arrival in the **temperate zone** for winter. The story was used to explain where the geese went in the summer before bird **migration** was understood (in fact, barnacle geese go to the Arctic to breed). A creative explanation!

BARNACLE GOOSE IN FRANCE -
DO YOU SEE ANY SIMILARITY?
© Nabok, Flickr

SEX-CHANGING CLOWNFISH

CLOWNFISH make their homes among the stinging tentacles of sea **anemones**. The **anemone** provides a safe home for the fish, offering it scraps of food and protection as no predators can enter the **anemone** without being badly stung (a sting to which only the clownfish is immune). In return, the clownfish keeps the **anemone** clean and healthy and chases away any **anemone**-eating fish. They live in the **anemone** as a male and female pair, alongside a few other non-breeding males (clownfish are all born male!). The female is the dominant fish, and the biggest. When she dies, her partner gains weight and becomes a female while the largest of the male non-breeders becomes her mate.

CLOWNFISH.
© PDTillman

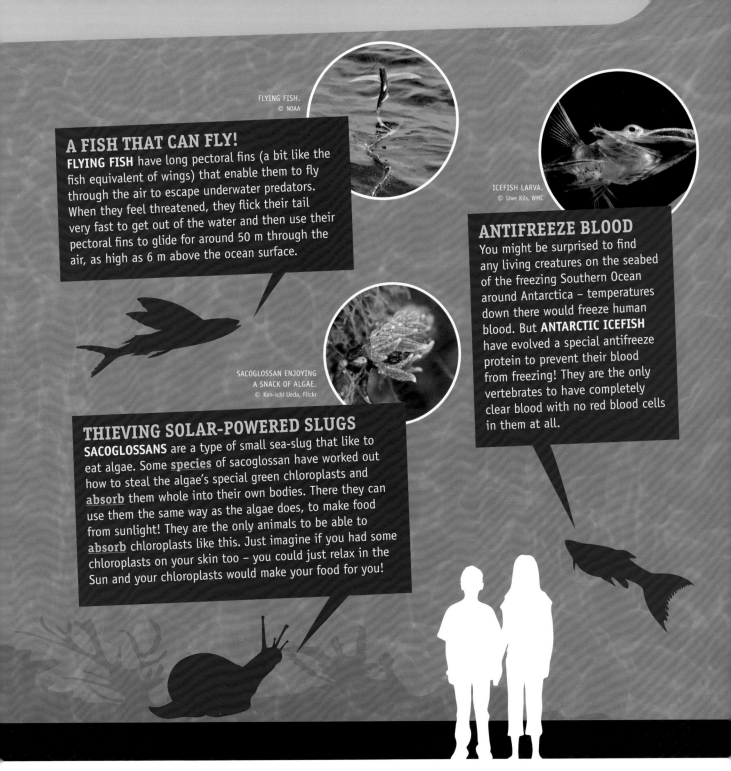

A FISH THAT CAN FLY!

FLYING FISH have long pectoral fins (a bit like the fish equivalent of wings) that enable them to fly through the air to escape underwater predators. When they feel threatened, they flick their tail very fast to get out of the water and then use their pectoral fins to glide for around 50 m through the air, as high as 6 m above the ocean surface.

FLYING FISH.
© NOAA

ICEFISH LARVA.
© Uwe Kils, WMC

ANTIFREEZE BLOOD

You might be surprised to find any living creatures on the seabed of the freezing Southern Ocean around Antarctica – temperatures down there would freeze human blood. But **ANTARCTIC ICEFISH** have evolved a special antifreeze protein to prevent their blood from freezing! They are the only vertebrates to have completely clear blood with no red blood cells in them at all.

SACOGLOSSAN ENJOYING
A SNACK OF ALGAE.
© Ken-ichi Ueda, Flickr

THIEVING SOLAR-POWERED SLUGS

SACOGLOSSANS are a type of small sea-slug that like to eat algae. Some <u>species</u> of sacoglossan have worked out how to steal the algae's special green chloroplasts and <u>absorb</u> them whole into their own bodies. There they can use them the same way as the algae does, to make food from sunlight! They are the only animals to be able to <u>absorb</u> chloroplasts like this. Just imagine if you had some chloroplasts on your skin too – you could just relax in the Sun and your chloroplasts would make your food for you!

CREEPY CREATURES

A SWARM OF HAGFISH.
© SERPENT Project

SLIMY SILK

The **HAGFISH** is very weird and unique in many ways. The <u>species</u> is thought to have barely changed for 300 million years, when it would have been one of the first animals to evolve a hard skull. It has no bones, can tie itself into knots, likes eating dead whales on the seabed and oozes lots and lots of slime. Some scientists are trying to use this slime to make a sort of silk for clothes! Anybody want a nice slime-silk shirt?

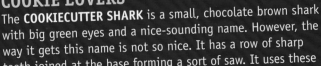

COOKIE LOVERS

The **COOKIECUTTER SHARK** is a small, chocolate brown shark with big green eyes and a nice-sounding name. However, the way it gets this name is not so nice. It has a row of sharp teeth joined at the base forming a sort of saw. It uses these teeth to cut out cookie-shaped chunks of flesh from its prey, which can include creatures from whales to humans to its big cousin the great white shark! It camouflages itself in the <u>water column</u> by glowing with a pale green light – this makes it blend in to the water better by hiding its shadow. Eerie!

COOKIECUTTER SHARK COMPARED TO THE SIZE OF A PENCIL.
© NOAA Observer Project

IRUKANDJI JELLYFISH QUEENSLAND, AUSTRALIA.
© GonwanaGirl

MINI KILLERS

IRUKANDJI JELLYFISH, along with the other box jellyfish, are the most poisonous animals alive. Their poison is 100 times stronger than a cobra's, and 1 000 times stronger than a tarantula's. Unlike the other box jellyfish, however, the Irukandjj is teeny tiny – only between 5 and 25 mm wide. Having said that, their four thread-thin tentacles can be up to a metre long. Being stung by one results in the dreaded Irukandji Syndrome: extremely severe pain and a sense of impending doom. Irukandjis are normally found in the waters around the north coast of Australia, but wait! <u>Climate change</u> is allowing the Irukandji to increase its range... Steer clear.

TONGUE EATING LOUSE
ON A SNAPPER FISH.
© Andy Heyward

TONGUE-TIED PARASITES

The **TONGUE-EATING LOUSE** is possibly one of the creepiest parasites out there. It is an isopod (related to pillbugs and woodlice you see on land) and it lives in the Caribbean Sea. It swims up to an unlucky fish, crawls into its mouth through the gill openings, eats the fish's tongue and attaches itself in its place! Sitting in the fish's mouth, it feeds on the fish's mucus and blood. The fish gets on with its life and uses the parasite just like a normal tongue. This is the only known case of a parasite functionally replacing a host's organ.

WANTED DEAD OR ALIVE

You are probably pretty familiar with what it's like to get a cold or the flu. You may also know that these common diseases are caused by microscopic particles called viruses. But did you know that viruses aren't really alive, and they aren't really dead either? They don't produce their own energy, and they rely on infecting 'fully-living' cells in order to reproduce themselves. In the ocean, **MARINE VIRUSES** are thought to infect everything from the tiniest bacteria to the biggest blue whale; they may even infect other viruses.

There are absoultely LOADS of them in the oceans – as in 100 million times more of them than there are stars in the whole universe! In fact, some even look as though they actually come from a distant galaxy, with their strange 3D hexagon-shaped heads and spindly spiderlike legs. Scientists are only just realizing how important viruses are to life (and death) on this planet. When these viruses infect bacteria, for example, they cause them to burst, releasing the <u>nutrients</u> the bacteria contain back into the ocean. These <u>nutrients</u> are then available for other <u>organisms</u> to take up.

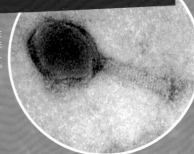

UNIDENTIFIED MARINE
VIRUS BELONGING TO THE
MYOVIRIDAE FAMILY.
© Mike Allen

CONTRIBUTORS & ORGANIZATIONS

LEARN MORE ABOUT THE PEOPLE WHO WROTE AND HELPED DEVELOP THIS BOOK AND ABOUT THE INSTITUTIONS THAT HAVE BEEN INVOLVED IN ITS PREPARATION.

The following annex contains information about the people and institutions that contributed to this Guide. They hope that you have found the Guide interesting and useful, but most of all, that you are now passionate about the world's fantastic ocean and will undertake your own actions to safeguard it.

CLOWNFISH IN AN ANEMONE - LOOKING AT YOU!
© q phia, Flickr

José Aguilar-Manjarrez is an Aquaculture Officer at FAO. His current responsibilities include planning, promoting, implementing and providing technical support to programmes and activities in the field of spatial planning for sustainable aquaculture development.

David Billett is a deep sea ecologist with an interest in the environmental management of the deep ocean. He is a Visiting Research Fellow at the National Oceanography Centre, Southampton, UK, and is a member of the Legal and Technical Commission at the International Seabed Authority.

Kelvin Boot is a Science Communicator at Plymouth Marine Laboratory, with a background in biology and geology and experience working with many science organizations including the BBC; his aim is to encourage understanding, appreciation and concern for the natural environment.

Jennifer Corriero is a social entrepreneur and youth engagement strategy consultant with a Masters in Environmental Studies from York University. She is co-founder and Executive Director of TakingITGlobal, and has been recognized by the World Economic Forum as a Young Global Leader.

Kelly-Marie Davidson has been the Communications Officer at Plymouth Marine Laboratory for the past seven years, having moved to the laboratory from a regional newspaper. She is currently in her second year of a three-year MSc course in Science Communication at the University of the West of England.

Matt Doggett is a professional marine biologist and wildlife photographer based in the UK. As a professional SCUBA diver, he has dived in locations around the world. He loves nothing more than opening people's eyes to the wonders of the seas through his photography. In 2012 Matt was Overall Winner of the British Wildlife Photography Awards.

Emily Donegan is a freelance writer and designer working for YUNGA. She holds a degree in Plant Sciences from Cambridge University and is strongly interested in sustainable living and ecology, but has always illustrated, painted and doodled in her spare time.

Annie Emery completed an internship with YUNGA and is currently studying Geography at the University of Cambridge. Within this course, she has studied ocean and coastal conservation issues, as well as the longer term impacts of climate change on these environments.

Helen Findlay is a biological oceanographer researching the impacts of climate change and ocean acidification on UK and Arctic marine systems.

Nicole Franz is a Fishery Planning Analyst at FAO, working primarily on small-scale fisheries policies and institutions, including on their socio-economic dimensions.

Alashiya Gordes is an environmental communicator with an MSc in Environmental Policy from Oxford University. She coordinates and edits YUNGA publications, supports FAO's climate change programmes and facilitates FAO's participation in various interagency groups on youth participation.

Caroline Hattam is an environmental economist at Plymouth Marine Laboratory. She works on projects that explore the importance of the ocean to human well-being and that aim to encourage the sustainable use and management of the marine environment.

Tara Hooper is an environmental economist at Plymouth Marine Laboratory and is the founder of the Marine Education Trust. She is particularly interested in marine renewable energy, tropical marine ecology and in tools for helping managers to make better decisions about marine conservation.

Frances Hopkins is a marine chemist at Plymouth Marine Laboratory where she studies the interactions between the surface ocean and the atmosphere, and how these processes are influenced by marine biology and chemistry.

Jennifer Lockett has a degree in Ecology from Plymouth University and several years' experience managing the Exe Estuary in Devon, a wetland of international conservation importance due to overwintering wildfowl.

Alessandro Lovatelli is a marine biologist with a specialization in marine aquaculture. He has a long working experience with FAO and other international development agencies on projects supporting the development of a sustainable mariculture industry through capacity development, technology transfer and policy development.

Ana M. Queirós is a marine ecologist at Plymouth Marine Laboratory, interested in understanding and mitigating human impacts on seabed species, habitats and processes.

Reuben Sessa is a Programme Officer at FAO developing and coordinating programmes on climate change. He is also FAO focal point for youth, coordinator of the YUNGA initiative and member of the Inter-Agency Network on Youth Development.

Jack Sewell has been involved in research and scientific interpretation at the Marine Biological Association in Plymouth for more than nine years. He has a degree in marine biology and coastal ecology and a MSc in Coastal and Ocean Policy.

Richard Shucksmith is a professional marine ecologist and environmental photojournalist living on the Shetland Isles. He has had a long life passion with watery habitats and the animals and plants that live and rely on these amazing ecosystems. He has won many awards for his work and was Overall Winner of the British Wildlife Photography Awards in 2011.

Doris Soto is a Senior Aquaculture Officer at FAO, leading activities under the Ecosystem Approach to Aquaculture (EAA) and promoting Climate-Smart Aquaculture practices at a global level.

Jogeir Toppe is a Fishery Industry Officer at FAO. He works on issues related to the role fish can play in improving food and nutrition security, and how we can improve handling, processing and preservation of fish products.

Christi Turner is an environmental reporter and youth environmental programme manager in Boulder, Colorado, and spent over six years in Madagascar working in conservation and sustainable development, most recently using media as a tool for youth engagement.

www.blueventures.org

Blue Ventures is an award-winning, science-led social enterprise that works with local communities to conserve threatened marine ecosystems and coastal livelihoods. They develop and scale innovative models for marine conservation within some of the poorest tropical coastal communities. Blue Ventures works in places where the ocean is vital to local people, cultures and economies, and where there is a fundamental need to support human development. The results of their work help them to innovate new ways to protect marine biodiversity that benefit coastal people everywhere.

www.cbd.int

The Convention on Biological Diversity (CBD) is an international agreement that commits governments to maintain the world's ecological sustainability through biodiversity conservation, the sustainable use of its components, and the fair and equitable sharing of the benefits arising from the use of genetic resources.

www.earthinfocus.com

Earth in Focus is a collaborative project by four professional ecologists and wildlife photographers with additional Associates. Their remit is to tell insightful ecological stories, engage people with science and the natural world and encourage them to explore and marvel at the wonders of our planet Earth.

www.fao.org

The Food and Agriculture Organization of the United Nations (FAO) leads international efforts to defeat hunger. FAO acts as a neutral forum where all nations meet as equals to negotiate agreements and debate policy. FAO is also a source of knowledge and information, helping countries to modernize and improve agriculture, forestry and fisheries practices and promoting good nutrition for all. FAO's Fisheries and Aquaculture Department in particular aims to strengthen global governance and the managerial and technical capacities of FAO's member states to improve the conservation and utilization of aquatic resources, hereby contributing to human wellbeing, food security, poverty alleviation and environmental sustainability.

http://ioc.unesco.org

The Intergovernmental Oceanographic Commission (IOC) of UNESCO is the United Nations body for ocean science, ocean observation systems, ocean data and information exchange, as well as ocean services, such as tsunami warning systems. Its mission is to promote international cooperation and to coordinate programmes in ocean research, services and capacity building; and to apply this knowledge to improve the management, sustainable development and protection of the marine environment and coastal areas.

www.mba.ac.uk

The Marine Biological Association is a learned society, undertaking world-leading marine biological science with a charitable remit to share this information with as wide an audience as possible. The MBA was founded in 1884 and in 1888 opened the Plymouth Laboratory at Citadel Hill. It has over 1 000 members worldwide and is home to the National Marine Biological Library and MarLIN (The Marine Life Information Network), and produces the Journal of the Marine Biological Association.

www.marineeducationtrust.org

The Marine Education Trust (MET) was established with the aim of engaging local communities in the sustainable management of their marine resources and enthusing young people to become advocates for marine conservation. MET's main areas of expertise are: developing innovative education resources; facilitating knowledge exchange; and supporting other small organisations in the delivery of environmental education and outreach activities.

www.noc.soton.ac.uk

The National Oceanography Centre (NOC), Southampton, is one of the top five oceanographic centres in the world for integrated ocean research and technology development covering all of the oceanographic disciplines. The deep seas group at NOC specialises in the ecology of the deep ocean including abyssal sediments, seamounts, mid ocean ridges and hydrothermal vent systems, all sites of current interest for the exploration of marine minerals.

www.TheOceanProject.org

The Ocean Project inspires action to protect our ocean. It helps aquariums and other visitor-serving organizations to engage with their millions of visitors effectively, fostering a more sustainable society. The Ocean Project's network of over 2 000 partner zoos, aquariums, museums (ZAMs) and other conservation organizations is the largest ever developed. The Ocean Project empower its partners with cutting-edge communications research, tools and resources that help ZAMs capture people's imaginations, create more engaged citizens and communities, and result in a significant impact. The Ocean Project is the lead coordinator of World Oceans Day.

www.pml.ac.uk

Plymouth Marine Laboratory is recognised worldwide for undertaking ground-breaking research in environmental science, both at the national and international level. Our research is based on the understanding that the ocean is vital to human existence, yet is being threatened by global change. Our scientists undertake observations and experiments, understand what these mean and turn resulting data and information into models to forecast what the impacts of the many pressures on the marine ecosystem will be. The knowledge which results from our research is shared by us with our many stakeholders, including policy makers, customers, fellow academics and the general public.

www.tigweb.org

TakingITGlobal is a non-profit organization with the aim of fostering cross-cultural dialogue, strengthening the capacity of youth as leaders, and increasing awareness and involvement in global issues through the use of technology.

www.wagggsworld.org

The World Association of Girl Guides and Girl Scouts (WAGGGS) is a worldwide movement providing non-formal education where girls and young women develop leadership and life skills through self-development, challenge and adventure. Girl Guides and Girl Scouts learn by doing. The association brings together Girl Guiding and Girl Scouting associations from 145 countries, reaching 10 million members around the globe.

ORGANI

www.WorldOceansDay.org

World Oceans Day is an opportunity to honour the world's oceans. No matter where we live, each of us is dependent on a healthy ocean for our survival. Recognized by the United Nations, World Oceans Day is held June 8th of every year. World Oceans Day serves as a global rallying point for raising awareness and promoting personal and community action in fun and positive ways, leading to a more aware, engaged, and sustainable society, and a healthier ocean and climate. Thanks to people like you, approximately 600 events were held to celebrate World Oceans Day in 2014. Let's make this year even better by holding great events, sharing, and spreading the word!

www.scout.org

The World Organization of the Scout Movement (WOSM) is an independent, worldwide, non-profit and non-partisan organization which serves the Scout Movement. Its purpose is to promote unity and the understanding of Scouting's purpose and principles while facilitating its expansion and development.

www.yunga-un.org

The Youth and United Nations Global Alliance (YUNGA) was created to allow children and young people to be involved and make a difference. Numerous partners, including UN agencies and civil society organizations, collaborate to develop initiatives, resources and opportunities for children and young people. YUNGA also acts as a gateway to allow children and youth to be involved in UN-related activities, such as the Millennium Development Goals (MDGs), food security, climate change, biodiversity and environmental sustainability.

PLEASE NOTE THAT THE INVOLVEMENT OF AN INSTITUTION OR INDIVIDUAL DOES NOT IMPLY ITS OR THEIR ENDORSEMENT OR AGREEMENT WITH THE CONTENT OF THIS GUIDE.

RELEVANT ORGANIZATIONS & CONVENTIONS

CRABBIE SEAL STRIKES A POSE.
© Jade Berman, Flickr

Convention for the Conservation of Antarctic Seals
An international agreement to protect, study and make sustainable use of seal species found in Antarctica.

Convention on the Conservation of Antarctic Marine Living Resources (CCAMLR)
An international agreement to protect the marine life of the Antarctic region.

Global Ocean Commission
An international body set up to formulate feasible recommendations to address the problems of overfishing, large-scale habitat and biodiversity loss, the lack of effective management and enforcement and deficiencies in high seas governance.

United Nations Educational, Scientific and Cultural Organization

Intergovernmental Oceanographic Commission

Intergovernmental Oceanographic Commission
The United Nations body for ocean science, ocean observation systems, ocean data and information exchange, as well as ocean services, such as tsunami warning systems. Its mission is to promote international cooperation and to coordinate programmes in ocean research, services and capacity building.

International Convention for the Regulation of Whaling
An international agreement to protect whales from overfishing, and which was the founding document for the International Whaling Commission.

International Maritime Organization
The United Nations specialized agency responsible for improving maritime safety and preventing pollution from ships.

International Seabed Authority
The organization responsible for controlling the extraction of mineral resources from the seabed in international waters, made up of representatives from different national governments.

TROUT KEPT FRESH ON ICE.
© Ann Wuyts, Flickr

Regional Fisheries Management Organizations
Regional fisheries management organizations are responsible for managing fish stocks in international waters, that is, fish stocks which are beyond the control of any national government.

United Nations Convention on the Law of the Sea
An international agreement that defines the rights and responsibilities of nations with respect to their use of the world's ocean.

BACKGROUND IMAGE
RICH SEALIFE.
© Comstock/Thinkstock

OCEAN WAVE.
© gdefon.com

GLOSSARY

Absorb: To take something up or retain it, for example, both the ocean and the atmosphere take up heat from the Sun's rays.

Acid, acidic: A substance that has a pH of less than 7 and releases hydrogen ions when dissolved in water.

Albedo: The proportion of incoming light from the Sun that is reflected by the Earth's surface.

Algal bloom: Fast-growing algae that can produce toxins which are harmful to other marine organisms and humans. These blooms vary in colour (purple, pink, red, green...). Though their exact cause is unknown, some appear to be the result of human activities including pollution and eutrophication.

Alluvial plain: A largely flat area of land created by a river depositing the sediment it carries over a long period of time.

Anemone: A soft-bodied sea creature with many stinging tentacles, and is related to corals.

Aquaculture: The farming of aquatic organisms including fish, crustaceans, molluscs and seaweed, usually in cages, ponds or, in the case of bivalves, on ropes or racks.

Atoll: A coral reef that encircles a lagoon partially or completely. An atoll forms when an underwater volcano sinks below the sea surface. Atolls are a special kind of fringing reef.

Atom: Everything in the world is made up of miniscule particles called 'atoms'. These particles are like small 'building blocks'. Different atoms combine to make up molecules of different substances.

Ballast water: Large volumes of water held in the tanks of large ships to keep them stable.

Barrier reef: A coral reef separated from the coast by a lagoon or channel.

Basin: A large underwater rock formation that dips down into a hollow (like a basin, hence the name).

Bays: A body of seawater that is partially enclosed by land, such as the Bay of Bengal, the Bay of Biscay and Baffin Bay.

Benthos: All the organisms living on and in the seabed (which is scientifically known as the 'benthic zone').

Biodiversity: The variety of all the different kinds of plant and animal life on earth, and the relationships between them.

Bio-irrigation: The process by which animals flush water (and other materials in the water) to and from the seabed.

Bioluminescent: An organism that produces its own light is said to be 'bioluminescent'.

Biomass: The mass of all living organisms in a given area.

Bioprospecting: Searching (prospecting) for plants, animals and other biological matter that can be used for products that can be sold commercially, such as medicine.

Bioturbation: The process by which animals move grains of sediment (and other materials) around.

Bivalves: Marine and fresh water molluscs whose bodies are enclosed inside two shells that are hinged together. They mainly feed by filtering particles out of the water.

Brackish: Water found where saltwater and fresh water mix together (e.g. at river mouths) causing the water's salinity to be higher than in fresh water but not as high as in seawater.

By-catch: Most fishers target specific kinds of fish, but are likely to catch other fish (or other species including dolphins, turtles and birds) unintentionally. These unintentionally caught species are known as by-catch.

Carbon dioxide: A gas made up of carbon and oxygen, which makes up less than one percent of the air. CO_2 is produced by animals and used by plants and trees. It can also be produced by human processes such as burning fossil fuels. CO$_2$ is a greenhouse gas and contributes to climate change.

Carbon sink: A reservoir that removes carbon dioxide from the atmosphere and stores it. A particular species, habitat or sediment can act as a carbon sink. The ocean as a whole is a very important carbon sink; forests and certain kinds of soils are other examples.

Cargo: The goods or produce transported by ships (or other forms of transport).

Carnivores: Carnivores are animals that gain all (or the vast majority) of their nutritional needs from eating other animals. Carnivore means 'meat eater' in Latin.

Challenger Deep: The deepest point in the ocean (at nearly 11 km below the surface) found at the southern end of the Marianas Trench in the Pacific ocean.

Chemosynthesis: A biological process that involves the use of inorganic substances such as methane and hydrogen sulphide as a source of energy to convert carbon molecules and nutrients into organic matter. It is an alternative to photosynthesis for producing food when no light is present (e.g. on the deep sea floor).

Climate: The long-term average, or overall picture, of the everyday weather experienced in a particular place.

Climate change: A long-lasting change in weather patterns that may occur over periods, lasting decades to millennia. It is caused by many factors including human activities, volcanic eruptions, changes in ocean currents and changes in the activity of the Sun.

Coastal squeeze: The term used to describe what happens to coastal habitats that are trapped between a fixed boundary on land (e.g. a sea wall or human settlements) and rising sea levels or increased storminess. The habitat is effectively 'squeezed' between the two forces and diminishes in quantity and or quality.

Coastal zone: The area where the land and sea meet and influence each other.

Cold seeps: These are found on the ocean floor where hydrogen sulphide, methane and other hydrocarbon fluids escape from the ocean floor. The animals found here use chemosynthesis to produce food. Cold seeps may also be known as gas seeps.

Colony: In ecology, this refers to a group of organisms of the same species living close together.

Colonize: In ecology, this refers to the process by which organisms become settled in a new area.

Condensation: The process by which gas or vapour cools and turns into a liquid.

Continental drift: The movement of Earth's continents as the tectonic plates that make up the planet's surface shift relative to one another.

Continental slope: The seaward border of the continental shelf, where the seabed drops sharply into the deep sea. The shallow areas of the seabed generally occur between the sea shore and the continental slope.

Convention: A Convention is a way in which something is usually done by a group of people. In international politics, international agreements are often called 'Conventions' (see pp.230-231 for a list of Conventions related to marine management).

Coral bleaching: Coral bleaching occurs when corals turn white because pressures have made them expel the microscopic, colourful algae living within their tissues. High water temperature is a major pressure, but high light intensity, low salinity, high acidity and pollutants also worsen the situation. Prolonged bleaching can kill the coral.

Corridors: In ecology, this refers to areas connecting different habitats: they can be thought of as transport routes that allow animals to move from one habitat to another.

Cosmopolitan: In ecology, this refers to species with a wide distribution, i.e. they can be found in many areas of the world. (The opposite of endemic species).

Crustacean: A group of mostly aquatic animals with hard shells and multiple legs. Crustaceans include crabs, lobsters, shrimps and barnacles.

Currents: Continuous and directed movements of water. Ocean currents are caused by the tides, wind and differences in the temperature and salinity of seawater.

Dead zones: These are areas of the ocean, often close to the coast, where little oxygen is found in the water and sediments making it difficult for marine life to live there. The number of dead zones in our oceans is growing. Also see eutrophication.

Debris: Discarded waste, the remains of something that has been destroyed.

Delta: A fan-shaped area of usually muddy soil at the mouth of a river that divides the river into smaller streams and channels before flowing into the sea.

Density: Density is the mass (or weight) of something relative to its volume. An object that is more dense has a greater mass for a given size. For example, a 1 cm³ piece of rock is more dense than a 1 cm³ piece of foam.

Desalination: The process by which salt is separated from water, so that fresh water is left behind. Desalinating water on a large enough scale to provide fresh water for everyday human use is expensive, as it requires special technology and a lot of energy.

Detritovores: Important organisms for decomposition: detritivores satisfy their nutritional needs by eating the dead bodies or debris of other animals and plants, and the waste products of other animals.

Developing country: A poor country that is trying to become more economically advanced. Developing countries tend to rely heavily on natural resources and subsistence farming or fishing (where farmers or fishers grow, raise or catch enough food only to feed their families, and rarely produce enough to sell to earn a living).

Diluted: Made thinner or weaker, for example by adding water.

Discards: Unwanted fish or other species that are accidently caught by fishers and thrown back into the sea. By-catch is often discarded.

Earthquake: A sudden violent shaking of the ground, typically causing great damage. Earthquakes occur when the Earth's crust moves, or due to volcanic action.

Ebb: The falling or outgoing tide.

Echinoderms: Living organisms found only in the marine environment. There are over 70 000 known living species of echinoderms in the ocean, including starfish, sea urchins and sea cucumbers.

Ecosystem: A community of living organisms (plants and animals) and non-living things (water, air, soil, rocks, etc.) interacting in a certain area. Ecosystems don't have a defined size: depending on the interactions you are interested in, an ecosystem can be as small as a puddle or as big as the entire ocean. Ultimately, the whole world is one big, very complex ecosystem.

Ecosystem services: The direct and indirect contributions of natural ecosystems to human wellbeing. These include the provision of food and raw materials, opportunities for recreation, emotional wellbeing and physical health. There are four types of ecosystem services: provisioning, regulating, cultural and supporting.

Eggs: Female reproductive cells, which are fertilized by male reproductive cells (sperm) to create young.

Endemic: A species that is native to a particular area or environment and not found naturally anywhere else.

Engineering species: Also known as 'ecosystem engineers', engineering species are organisms that modify the physical and chemical characteristics of the environment around them. This includes anything from a tree root that surfaces and changes the shape of your garden, to the complex effects that living organisms have on the chemistry of the seabed.

Equator: The equator is the line around the Earth at 0° latitude, where the Sun is directly overhead at noon on the two days a year where day and night have exactly the same length (these days are called equinoxes and occur around 21 March and 21 September).

Erosion: Erosion means 'wearing down'. Rocks and soils are eroded when they are picked up or moved by rain, running water, waves, ice, gravity, or other natural or human agents. Also see weathering.

Estuary: A type of river mouth, into which the sea enters causing a mixing of fresh water and seawater.

Eutrophication: Caused by the presence of excessive levels of nutrients, eutrophication often occurs in coastal waters. It results in the fast growth of phytoplankton and other marine algae which can contribute to the creation of dead zones.

Evaporate/evaporation: The process by which heat turns a liquid substance into vapour.

Exclusive Economic Zone: The area of sea extending up to 200 nautical miles (370 km) from a country's coastline, over which that country has special rights to exploit the marine resources (such as fish, fossil fuel supplies and minerals).

Extinction: When a particular species is no longer alive on Earth, it is 'extinct'.

First-year ice: Newly formed sea ice that hasn't existed for longer than a year.

Fjord: A steep-sided, narrow inlet that was carved over the centuries by a glacier. (A glacier is a huge, long-lasting 'river' of ice.)

Flood: When talking about the tides, 'flood' refers to the rising or incoming tide (the opposite of ebb). More generally, flooding means an area of land has been covered by water (e.g. due to heavy rain or because rivers or lakes have flooded the land around them).

Food chain: The links between organisms, showing what eats what. A food chain shows how energy passes between individuals, starting with primary producers (plants) all the way up to carnivores and detritivores. For instance, in coastal areas, small organisms eat very small seaweeds and bacteria; these are then eaten by larger animals like fish, which in turn are themselves eaten by even larger fish, birds and mammals.

Food web: A more complicated version of a food chain, showing that more than one animal may have the same food source, meaning that different food chains are interconnected.

Fossil: The preserved remains of an ancient animal or plant.

Fossil fuels: Fossil fuels form over millions of years from prehistoric plant or animal remains. The three fossil fuels are coal, oil and natural gas. When we burn fossil fuels to fuel vehicles or generate energy, the greenhouse gas carbon dioxide is released into the atmosphere, contributing to climate change.

Fresh water: Naturally occurring water that is not salty (e.g. in rivers, lakes and groundwater).

Fringing reef: A relatively young coral reef found close to the coast. Fringing reefs are one of the two main types of coral reef, alongside barrier reefs. Also see atoll.

Geological: Related to the rocks of the Earth's crust.

Gravity: The force of attraction between two objects. This may also be referred to as gravitational pull.

Greenhouse gas: These are gases in the atmosphere that can absorb and emit (or radiate) heat. They include water vapour, carbon dioxide, methane, nitrous oxides and ozone. Human activities like industrial production, energy production and transportation have increased the levels of greenhouse gases in the atmosphere to such an extent, that the Earth's temperature is starting to rise: this is known as climate change.

Groundwater: Water located beneath the Earth's surface, often feeding springs and wells. This is the Earth's biggest storehouse of drinkable water.

Gulf: A large area of seawater partially enclosed by land. Gulfs are usually much bigger than bays. Examples include the Gulf of Mexico, the Gulf of Aden and the Gulf of Bothnia.

Gyres: These are large systems of rotating ocean currents, usually associated with wind driven currents. There are five major gyres: one in the north Atlantic, one in the south Atlantic, one in the north Pacific, one in the south Pacific and one in the Indian Ocean.

Habitat: The local environment within an ecosystem in which an organism usually lives. The attractiveness of particular habitats to particular creatures depends on seabed type (sand, mud or rock for example), factors such as water temperature and salinity, and the presence of certain types of marine life, particularly those that form living reefs.

Herbivores: Animals that only eat plants, algae and photosynthesizing bacteria.

High seas: The area beyond the Exclusive Economic Zone of a particular country. Also known as 'international waters', no single country has legal rights over the high seas.

High tide: The highest level on the shore reached by the sea on a particular day (also called 'high water').

Hurricane: An extremely intense tropical storm that forms out in the ocean producing very strong winds and heavy rain. In different parts of the world, hurricanes are called typhoons or tropical cyclones.

Hydrothermal vents: Openings on the ocean floor where naturally heated water escapes, often associated with volcanic activity.

Hypoxia: This occurs in ocean environments when the level of dissolved oxygen in seawater is reduced so much that it can no longer support marine life. In extreme cases of hypoxia, the area becomes a dead zone.

Indigenous: Something that originates from or occurs naturally in a place, rather than having been introduced (e.g. by human activity).

Inland seas: Land-locked bodies of water or salt lakes that show characteristics similar to seas.

Intertidal zone: The area of the coastline between the levels of low tide and high tide: the intertidal zone is exposed as the tide goes out and covered with water again as the tide comes in again.

Invasive species: Animals, plants and other species that have been introduced to an area from elsewhere, either by accident or on purpose, and negatively affect the native habitat by out-competing native species.

Invertebrate: An animal which does not have a backbone.

Juvenile: A young animal that is not fully grown.

Larvae: Animals that have just hatched from their eggs. The larvae of many marine animals (including crabs, molluscs and worms) look very different from their adult form.

Latitude: A measure of the distance north or south of the equator.

Low tide: The lowest level on the shore reached by the sea on a particular day (also called 'low water').

Mangroves: Salt-tolerant trees found in the coastal zones of tropical areas. The term 'mangrove' can refer to an individual tree or to a whole forest.

Marianas Trench: The deepest area of the ocean, found in the western Pacific Ocean.

Mariculture: Marine aquaculture – that is, the farming of marine organisms on shore or in coastal waters.

Marine parks: Highly protected areas of the ocean in which human activities are strictly controlled to avoid damage to the natural ecosystem. Marine parks are a kind of Marine Protected Area.

Marine Protected Area (MPA): A protected area in the marine environment in which some or all human activities are restricted to help protect marine habitats and cultural or historical resources found there.

Marine resources: Useful things provided by the ocean, such as fish, minerals or even the opportunity for recreation.

Marine snow: Dead bodies and plant or animal waste that fall from the upper ocean to the deep sea.

Marsh flat: A low-lying area of saltmarsh.

Methane: A chemical compound (usually a gas) made from carbon and hydrogen. Methane is an important fuel for cooking and heating, but also a significant greenhouse gas.

Microbe: A microscopic organism.

Migrate, migration: 'Migration' means to 'move' in Latin. Migratory animals travel a long distance on a regular basis (e.g. some whales migrate from feeding grounds in the Arctic to breeding grounds in the Caribbean).

Molecule: When individual atoms stick together, they make up small clusters called 'molecules'.

Different molecules make up different substances. Water, for example, is made up of molecules which contain two hydrogen (H) atoms and one oxygen (O) atom, which is why water's scientific name is H_2O. An oxygen molecule is made up of two oxygen atoms, and is called O_2.

Molluscs: Animals without a backbone ('invertebrates') including snails, squids and octopus. About 23 percent of all named marine organisms are molluscs.

Monsoon: The period of heavy rain that occurs in summer in certain tropical and subtropical areas. The monsoon is caused by seasonal winds blowing from the cool sea onto the warm land.

Multi-year ice: Polar ice that has existed for longer than one summer (also see first-year ice).

Natural resources: Natural resources are useful materials found in the natural environment around us. Water, soil, wood or rocks are examples of natural resources we rely on to survive. For example, we need water for drinking, water and soil for growing food, wood for fuel, making paper and furniture, and wood and rocks for building materials. And those are only a few of the uses we can put those resources to! Can you think of more?

Neap tides: Neap tides are tides with a less extreme tidal range than spring tides. They occur when the Moon is in its first or third quarter.

Northern hemisphere: The area of the world to the north of the equator (in which Europe, North America, most of Asia and much of Africa are located).

Nutrients: Chemicals that animals and plants need to live and grow.

Ocean acidification: The increased acidity (or decreased pH) observed in the ocean as a result of its rapid uptake of carbon dioxide over the last century.

Organic matter: Substances that have a biological origin (i.e. living plants and animals, as well as debris and nutrients from dead and decomposing plants and animals).

Organism: A living creature, like a plant, animal or microorganism.

Overfishing: To decrease fish stocks by too much fishing.

Perennial: Lasting all year round and potentially, for many years (e.g. the ice cover that doesn't melt in summer is called perennial).

Periscope: An instrument that uses a set of lenses, mirrors or prisms which are set inside a casing to allow us to observe a concealed place. For instance, submarines use periscopes to observe events taking place above the water without revealing the submarine's position underwater.

pH: A measure of how acid (low pH) or alkaline (high pH) a substance is.

Photosynthesis: A biological process found in green plants and algae. Photosynthesis uses light as an energy source to convert carbon dioxide and water into a source of food (sugars and other useful chemicals). The term comes from Ancient Greek: 'photo' means 'light' and 'synthesis' means 'putting together'.

Phytoplankton: Small, microscopic marine plants that drift with the ocean currents. They live in the upper layers of the ocean and use photosynthesis to produce food.

Plankton: Microscopic marine plants and animals that drift with the ocean currents.

Polar regions: Together, the Arctic (the region surrounding the North Pole, which has a latitude of 90 °N) and the Antarctic (the region surrounding the South Pole, which has a latitude of 90 °S) are known as the 'polar regions' or simply, the 'poles'.

Polyps: In marine ecology, polyps are individual animals within the colonies that make up corals and anemones.

Primary producers: Organisms that can photosynthesize (i.e. plants and algae). Primary producers are the basis for all food chains.

Prism: A transparent object which makes light passing through it change direction ('refract').

Quota: A limit to something, in this case, the maximum amount of fish that a fisher is allowed to catch. Fishing quotas need to be enforced in order to protect fish stocks from overfishing.

Reef: A solid underwater structure that may be formed from rock, by marine creatures (such as corals or oysters), or deliberately or accidentally by humans (e.g. harbour pilings or sunken ships).

Renewable energy: Energy powered by renewable resources which can be replaced or replenished, either by natural processes or human action. Wind, water and solar energy are examples of renewable forms of energy.

Ria: A drowned river valley.

Rip currents: These are narrow, fast moving flows of water that travel away from the coast. They can occur at any beach where waves break.

Rocky shore: A beach characterised by the presence of rock platforms or large boulders.

Run-off: Substances (usually soil, chemicals or other pollutants) that have been washed off the land by rain and carried into rivers and the sea.

Salinity: 'Saline' is another word for 'salty'. 'Salinity' refers to the amount of salt dissolved within seawater (its 'concentration'). Seawater naturally has a high salinity, while fresh water is not very saline.

Saltmarsh: An expanse of highly salt-tolerant plants that are found at the top of the shore (usually behind mudflats) and which are sometimes submerged by high tides.

Sand bar: A ridge of sand, often at the mouth of a river or estuary, shaped by currents and waves.

Satellite: A piece of equipment humans send into space to orbit the Earth, usually in order to collect data or provide communications services.

Sea: A large body of salt water connected to an ocean. Often the word 'sea' is used interchangeably with the word 'ocean'.

Seafarers: People who work at sea.

Sea ice extent: The extent to which an area is covered by sea ice. Regions are defined either as 'ice-covered' or 'not ice-covered'.

Seamount: An underwater mountain that does not break through the surface of the sea.

Seawater: The water found in an ocean or the sea. It differs from fresh water because of the concentration of dissolved salts found in it.

Sediment: Different types of sand, mud and rocks which are carried and eventually deposited by wind, water or ice. For example, the seabed is made of deposited sediment, as are underwater structures like sand bars.

Sediment profiler imager: An object that uses a large prism to allow us to study and observe what's going on inside the seabed.

Sediment shores: Beaches made up of soft sediments such as sand or mud.

Solar energy: Energy from the Sun.

Southern hemisphere: The area of the world to the south of the equator (in which most of South America, Southern Africa, Australia and Antarctica are located).

Species: A group of similar organisms which are able to breed together and produce healthy offspring that are able to produce young themselves.

Sponge: Sponges are living organisms with bodies full of pores and channels that allow water to flow through them, from which they absorb the nutrients they need to survive. They do not have a nervous, digestive or circulatory system. In the deep sea, clusters of sponges may form dense mats known as 'sponge fields'.

Spring tides: Tides that produce higher high tides and lower low tides than average and occur when the Moon is new or full (in its second or fourth quarter).

Stock: The available amount of something. In this context, the amount of fish available in the ocean.

Storm surges: Caused by high winds, storm surges cause a rise in seawater resulting in tides that are higher than usual and may flood the coast.

Straits: Narrow channels of water that connect two larger bodies of water, for example the Straits of Gibraltar (which connect the Mediterranean Sea with the Atlantic Ocean) or the Bering Straits between Alaska and Siberia (which connect the Pacific Ocean with the Arctic Ocean).

Subduction: The process by which a tectonic plate moves underneath another and is pushed downwards into the Earth's mantle.

Subtropical: Relating to the subtropics.

Subtropics: The regions between the tropics and the temperate zones.

Superheated: A liquid that is heated to a temperature above its boiling point, but still remains liquid, rather than turning into a gas.

Surface currents: Wind-driven currents that form within the top 400 m of the ocean surface.

Sustainability: The careful use of natural resources by humans, making sure that this use meets our needs without damaging the environment (i.e. that it can continue to support plant, animal and human life). Making sure that our actions are sustainable means that future generations will be able to live well, too.

Sustainably sourced: This refers to products that are produced with environmental and social impacts in mind. For example, sustainably sourced fish is caught or raised using methods that do not exploit the ocean environment, fishers or fish farmers, or threaten fish stocks.

Tectonic plates: Huge areas of rock underneath the continents and ocean that make up the Earth's crust. Tectonic plates move very slowly (about 2-5 cm each year), over the surface of the Earth, grinding into, sliding beneath or moving away from each other. Also see earthquake, subduction and volcano.

Temperate zones: The areas between the tropics and the polar regions where the temperatures are relatively moderate with few extremes in winter and summer.

Terrestrial: Relating to the land or Earth as a whole ('terra' means 'earth' in Latin).

Territorial waters: The area of sea (including the seabed) that is under the complete control of a national government. Territorial waters usually extend for 12 nautical miles (22 km) from the country's low tide mark. The UN Convention on the Law of the Sea declares that foreign vessels are allowed 'innocent passage' through territorial waters, meaning they may have to comply with national restrictions and must be peaceful.

Thermohaline Circulation: 'Thermo' refers to temperature, 'haline' refers to salinity, and 'circulation' refers to movement. Putting all this together: we call the flow of ocean water caused by changes in water density resulting from changes in water temperature and salinity levels the 'Thermohaline Circulation' (THC).

Tidal barrage: A dam built across an estuary to control the flow of the tide, using the tide to drive turbines which generate electricity.

Tidal bore: A wave formed by the incoming tide that progresses into an estuary or bay in areas with a high tidal range.

Tidal current turbine: A device like an underwater windmill, which generates electricity as its blades are forced to turn by the speed of water flowing past it as the tide ebbs and floods.

Tidal cycle: The movement of the sea over the course of several hours, as it moves in from low tide, reaches high tide and then falls back to low tide.

Tidal range: The difference between the highest and lowest tides in an area.

Tides: The rise and fall of the sea due to the gravitational pull of the Moon and Sun and the turning of the Earth. Most places see two high and low tides per day. See gravity.

Transition zone: An area where two different areas or features meet, in which some characteristics of both can be found (e.g. river mouths are transition zones, because that's where fresh water from the river and salty seawater meet and mix).

Trawling: Fishing by towing large nets through the water behind one or several boats. Similarly, in bottom trawling nets are dragged along the seabed rather than through the water.

Treaty: A formally concluded (ratified) agreement between countries.

Tropic of Cancer: The line at about 23.5 °N, where the Sun is directly overhead at noon on the June solstice (around 21 June). That is the day of the year on which the Sun is at its highest in the sky when considered from the North Pole.

Tropic of Capricorn: The line at about 23.5 °S, where the Sun is directly overhead at noon on the December solstice (around 21 December). That is the day of the year on which the Sun is at its lowest in the sky when considered from the North Pole.

Tropical: Relating to the tropics.

Tropics: The areas around the equator, which have a very warm climate and about 12 hours of daylight (and 12 hours of darkness) throughout the year. The tropics extend north to the Tropic of Cancer and south to the Tropic of Capricorn.

Tsunamis: Extremely powerful waves caused by changes on the seabed including earthquakes, volcanic eruptions and underwater landslides.

Vapour: When liquid is heated, it evaporates and turns into a gas: this is known as 'vapour'.

Vent: A crack in the seabed, through which heat, gases and liquids can escape.

Volcano: A place (usually a mountain) over a break in the Earth's crust, through which molten rock, ash and gas are sometimes expelled. The process of expelling molten rock is called a volcanic eruption.

Water column: An imagined vertical cross-section of the sea extending from the seabed to the water's surface.

Water cycle: The water cycle describes the movement of water from the sea to the atmosphere to the land and back to the sea. It also describes the changes in state of the water from solid to liquid to vapour.

Weather: The outdoor conditions experienced on an hour-to-hour or day-to-day basis in a particular place, including the cloud cover, rainfall, air temperature, air pressure, wind and humidity (the amount of water vapour in the air).

Weathering: The wearing away of a material or substance such as rock or soil due to natural factors (like wind, rain, the tides, or growing tree roots) or human factors (like chemical pollution). Unlike erosion, weathering takes place without the material being moved.

Zooplankton: Microscopic marine animals that float with the ocean currents. Some zooplankton spend all of their lives as plankton, but others only spend their young (juvenile) stage as plankton, developing into larger adult phases (e.g. jellyfish and other fish species).

Zooxanthellae: Microscopic algae living within the tissues of reef-building warm water corals.

YOUR NOTES